Question Everything

132 science questions – and their unexpected answers

NewScientist

Question Everything

132 science questions – and their unexpected answers

More questions and answers
from the popular 'Last Word' column

edited by Mick O'Hare

First published in Great Britain in 2014 by
Profile Books Ltd
3A Exmouth House
Pine Street
Exmouth Market
London EC1R 0JH
www.profilebooks.com

10 9 8 7 6 5 4 3 2 1

A CIP catalogue record for this book is available from the British Library.

ISBN (paperback): 978 1 78125 164 5
ISBN (book club hardback): 978 1 78125 468 4
eISBN: 978 1 84765 984 2

Text design by Sue Lamble
Typeset in Palatino by MacGuru Ltd
info@macguru.org.uk

Printed and bound by
CPI Group (UK) Ltd, Croydon, CR0 4YY

Contents

Introduction

The Last Word column in *New Scientist* magazine – the publication in which most of the questions and answers in this book began life – has traditionally devoted itself to the trivial. That's why we know why hair turns grey, why snot is green, and how fat you need to be in order to be bulletproof. And it's why this book's predecessors have proved to be such a hit with readers. It's our stock-in-trade: finding out how and why the things surrounding our everyday lives do what they do.

But every so often we look at the bigger picture. We'll get a question about stuff in the sky, stuff in the outer reaches of the universe or stuff at the atomic level. Not so much trivia as the stories behind why our planet, our solar system and our universe are like they are. Just why is the night sky black even though it's full of stars pumping out more light than our sun? Why does Earth rotate? And why is the Large Hadron Collider so, er, large? You can find out all about these and other mega-quandaries in the pages that follow.

Obviously, we also have lots of the usual stuff in here about our bodies, our dinners and the various living things we share the planet with, to keep regular readers happy. You can find out why your legs feel wobbly when you are standing on the top of a cliff or how flies can fly into a window pane without damaging themselves. And there's a lot about booze, of course. Why does ice in whisky create such lovely swirling patterns? Does drinking absinthe make you hallucinate? As we say on the cover: 'Question Everything'.

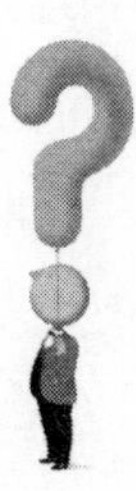

But this time we set out to make people think a little more about our universe and what's in it, and the physics, chemistry and biology that keep it all on the move.

So, with a big bang, here we go... and remember, you can follow the Last Word online at www.newscientist.com/lastword or in print by buying the weekly *New Scientist* magazine. Questions (and answers) always welcome!

Mick O'Hare

Special thanks – in no particular order – to the following: Melanie Green, the *New Scientist* subs and art teams, Jeremy Webb, Beverley De Valmency, Paul Forty, Andrew Franklin, Eleanor Harris, Drew Jerrison, Nick Heidfeld, Harold Wagstaff, Felipe Massa, Sally and Thomas.

1 Earth

In a spin

What makes Earth rotate?

R. J. Isaacs
Barnet, Hertfordshire, UK

Earth rotates simply because it has not yet stopped moving. The solar system, and indeed the galaxy, were formed by the condensation of a rotating mass of gas. Conservation of angular momentum meant that any bodies formed from the gas would themselves be rotating. As frictional and other forces in space are very small, rotating bodies, including Earth, slow only very gradually. The moon, a much smaller and lighter body, has effectively already stopped rotating because of the gravitational drag exerted by Earth, and now always keeps the same face turned towards us.

Glyn Williams
Derby, UK

Although Glyn Williams's explanation of Earth's rotation is correct, his assertion that 'the moon … has effectively already stopped rotating' could be misleading. The moon does rotate. The reason why it presents the same face to us is that its period of rotation is the same as its period of revolution around Earth. This equality is the result of tidal friction. If the moon did not rotate, any line through it, parallel to the orbital plane, would keep the same direction in space and the moon

would show us its far side during a complete revolution, as one can easily convince oneself by a simple drawing on paper.

D. S. Paransis
Department of Geophysics
Luleå University of Technology
Sweden

Early days

The shortest day of the year in the northern hemisphere occurs on 21 or 22 December, yet the earliest sunset is on 13 or 14 December and the latest sunrise occurs a similar number of days after the shortest day. Why is this?

John Walker and Alan Whittle
Manchester, UK

Two properties of Earth's orbital motion around the sun give rise to the curious disparity between the dates of earliest sunset, shortest day (winter solstice) and latest sunrise. These are the eccentricity of Earth's orbit, and the tilt of its equator to its orbital plane.

The combined effect of these is to vary the length of the day throughout the year. In some months, the interval between noons on successive days is slightly greater than 24 hours, while in other months it is slightly less. The differences cancel one another out over the course of a year.

In December, near the northern winter solstice, the interval between successive noons is about 30 seconds less than 24 hours. Because this difference is greater than the daily change in sunrise and sunset times, it becomes the dominant effect, causing the observed separation of the dates of earliest sunset and latest sunrise.

Fred Watson
Coonabarabran, New South Wales, Australia

The main effect of this shifting of latest sunrise to several days after the winter solstice (the shortest day of the year) is an oscillation of the time of day when the sun reaches its highest elevation. The time oscillates about noon in a sine wave with an amplitude of 8.8 minutes at latitude 45° and with a period of 6 months. At the solstices and equinoxes, highest elevation is at noon.

Each day, sunrises and sunsets occur equally before and after the time of the sun's highest elevation. If we call those intervals morning and afternoon, then at the winter solstice, the length of morning is at its minimum and so changes very little for a few days. But although the time of the sun's highest elevation is noon at the solstice, that time gets later quite quickly and shifts the sunrise so that it falls later for a few days. After those few days, morning begins to get longer and overwhelm the 8.8-minute sine wave so we get earlier sunrises again.

Why does the time of the sun's highest elevation oscillate around noon? Mainly because Earth's axis is tilted with respect to its orbit around the sun. Fix a frame of reference at the centre of Earth, aligned with the axis of Earth, but rotating uniformly around that axis only once per year, not once per day. In that frame, the sun generally goes up and down by 23 degrees. But it also goes side to side a little, and this gives the oscillation in time of the sun's highest elevation each day around noon.

Terry Watts
New Jersey, US

Midsummer misnomer

In the northern hemisphere the sun reaches its highest point in the sky on (or very close to) 21 June every year. Yet the warmest months tend to be July or August. Why is that? Similarly, on 21 December the sun reaches its lowest point but the coldest months are usually January and February. Can somebody explain this?

Wolfgang Wild
Granada, Spain

Earth has a certain heat capacity, which leads to a thermal time lag. Therefore, when a hemisphere is experiencing its longest day it is still warming up, and will not reach its warmest until a few weeks later. Similarly, on the shortest day it is still cooling down and will not reach its coolest until a few weeks later.

Aidan Westwood and Stephen Collins
University of Leeds, UK

Although the northern hemisphere receives most sunlight at the end of June, it is rather like heating up a room with a gas fire. The fire gets hot very quickly, but it takes a while for the air in the room to heat up. The same applies to the atmosphere of Earth. Likewise, the room does not get cold the instant the fire is turned off – the air gradually cools in the same way the air in Earth's atmosphere does.

Ian Hedley
Poole, Dorset, UK

Not early to rise

Being very keen to see the return of the sun after the northern hemisphere winter, I began checking the sunrise and sunset times to see how the interval between them increases each day. The sunset time increases by more than one minute per day but the sunrise time is getting earlier by considerably less than one minute each day. Why is there asymmetry? I suspect it has got something to do with London's latitude, but what?

Kate Lee
London, UK

Your correspondent has discovered a quirk of astronomy usually termed the 'equation of time'. I was wondering recently whether day length really changes faster at some times of the year and slower at others. I plugged sunrise and sunset times into a spreadsheet and the resulting graph shows that the day length stays briefly but depressingly static in midwinter, then changes rapidly through spring into another short period of near stasis in summer.

But then I noticed the oddly different behaviour of sunrise and sunset times: the two curves are not symmetrical. So I found the midpoint of the two, which gives the apparent (or true) solar time of noon. When you plot the difference between this time and the mean (or clock) time of noon across the year, you discover a complex curve in a second graph, which is a visual representation of the equation of time. Sunrise and sunset are symmetrical about this line.

Steve Head
Cholsey, Wallingford, Oxon, UK

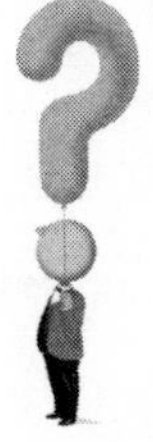

The asymmetry in the rates of change of sunrise and sunset arises from the nature of Earth's orbit around the sun, and is caused by variations in the length of the solar day – the time between solar noons on successive days – throughout the year. Sunrise and sunset are essentially symmetrical about solar noon, but solar noon is not always at clock noon. The sources of this variation are, firstly, Earth's elliptical orbit around the sun, and, secondly, the 23.5-degree inclination of Earth's rotational axis to the axis of its orbit around the sun.

From solar noon one day to solar noon the next, Earth not only has to turn through 360 degrees, but also through about 1 more degree to compensate for the movement along its orbital path during that time. While Earth reliably turns through 360 degrees once every 23 hours 56 minutes, regardless of the time of year, it is the variation in how much further Earth has to turn to complete the solar day that gives rise to the varying solar day lengths.

Earth speeds up as it approaches the perihelion of its elliptical orbit, the point of closest approach to the sun, and slows down as it approaches the aphelion, the point furthest from the sun. The increased speed at the perihelion, together with the shorter distance to the sun, means the angle swept out by Earth about the sun every day is greater near the perihelion than near the aphelion. So more rotation is needed to complete a solar day near the perihelion, causing the solar day to lengthen. This factor generates a sine wave-like variation in the length of the solar day with a period of one year.

The second source of the variation is more difficult to visualise. At the summer and winter solstices, the plane joining Earth's rotational axis to the centre of the sun is perpendicular to the plane of Earth's orbit, and a point on the equator has the least distance to travel between one noon and the next, so the solar day is at its shortest. By contrast, at the equinoxes, Earth's axis is tilted towards or away from the orbital plane. Now, a point on the equator must travel further between consecutive noons, and a solar day is longer than a solar day at the solstice by about 22 seconds. This mechanism operates at other latitudes, which causes a variation in the solar day length with two peaks and two troughs a year.

These two sources of variation together create an intricate pattern of solar day length. While the changes from day to day are small, they are cumulative and can lead to marked differences between solar time and clock time during a year. They are the bane of sundial makers and are expressed in the equation of time, which shows a pattern resembling a sine wave of a six-month period superimposed on a sine wave of a one-year period. The time difference ranges from about 14 minutes negative to more than 16 minutes positive, with the steepest slope, of more than 20 seconds a day, occurring in December. It is the shifting solar noon added to what would otherwise be a symmetrical movement of sunrise and sunset that produces the asymmetry.

Ian Vickers
Mosman, New South Wales, Australia

There are several websites where you can find out more about this. Try: www.analemma.com and www.cso.caltech.edu/outreach/log/NIGHT%20and%20DAY.pdf – Ed.

Long hours

Do all points on Earth receive an equal number of daylight hours over the course of a year? I realise that the intensity of the sun's rays will differ considerably depending on your latitude, but does a person living in Alaska see the sun for as long as somebody living in Ecuador over the course of a year?

Tami Shinn
By email, no address supplied

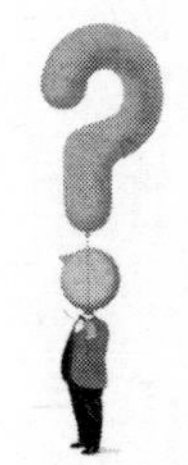

The short answer is that all points on the globe do receive the same amount of sunlight, because any portion of Earth that points towards the sun in summer will point away in winter, so the extra daylight in summer precisely cancels the lack of daylight in winter. However, this is true only for a simple spherical body in a circular orbit about the sun.

First, the orbit of Earth is not circular. According to Kepler's third law, Earth travels faster when it is closer to the sun than when it is further away.

Because Earth is closest to the sun in early January, it moves fastest during the northern hemisphere's winter months. This can be seen by the fact that the time from the autumn to the spring equinox is about three days shorter than the time from the spring to the autumn equinox.

Because Earth has more daylight during the northern hemisphere's summer and spends more time on the 'summer side' of the equinoxes, the northern hemisphere receives slightly more daylight than the southern. This amounts to about 6 hours of additional daylight per year at 50° north, with higher latitudes receiving even more daylight.

The second major effect is the refraction of sunlight by the atmosphere. Because of this, we can still see the sun after it has sunk below the horizon. Typically, sunset appears to happen when the sun is actually about half a degree below the horizon. Along the equator, the difference amounts to about 4 minutes per day when daylight is more than 50 per cent of the day.

While this gives some extra daylight to everyone wherever they live, it gives more daylight in areas where the path of the sun makes a shallow angle with the horizon.

The shallow angle means that it takes the sun longer to reach half a degree below the horizon. So higher latitudes (both north and south) get the most additional daylight. At 50° latitude, this is up to 8 minutes more per day when daylight is more than 50 per cent of the day – or about 36 hours per year on average.

Barry Spletzer
Albuquerque, New Mexico, US

On the pull

If you could journey to the centre of Earth, what would the sensation of gravity be at various points on the way down, and at the centre?

Charles Wright
Vancouver, Canada

This problem piqued the curiosity of no less a physicist than Isaac Newton, who of course solved it in his *Principia* (Book 1, theorem 33). If you are at the centre of Earth you are pulled equally in all directions, so you are in fact weightless. Higher up, at radius R from the centre, Newton found that the attractions of the materials in the hollow spherical shell of radius greater than R will all cancel one another out – a beautiful mathematical consequence of the fact that gravity decreases as the square of the distance between the objects increases. You feel only the pull of the mass in the sphere below you.

Newton showed that Earth's combined pull is simply proportional to the inverse square of the distance R from the centre. The mass of this sphere is proportional to its volume, that is, R^3. So, the weight you would feel, if you were foolhardy enough to descend through a homogeneous planet, would decrease in direct proportion to R^3/R^2 (which is equal to R) as you moved inwards, reaching zero at the centre.

In fact, the central parts of Earth – mostly dense iron – are much more massive than the outer parts, so your weight would decrease a bit more gradually at first and more rapidly as you penetrated the core.

Spencer Weart
Maryland, US

As you descended you would find the pull of gravity continuously decreasing as the centre of Earth approached, and at the centre you would be almost completely weightless, but not quite.

First, suppose Earth is perfectly spherical (even though it isn't quite). This would mean that at any interior point all the matter lying outside your position (that is, towards the surface of Earth) makes no contribution to the pull of gravity – it all cancels out to zero – leaving just the pull from the matter inside your position (all the matter between you and Earth's centre). So gravity gets even weaker as you descend, and would be zero at the centre.

However, Earth is slightly pear-shaped. Think of this as an extra 'ring' of matter on the surface just south of the equator. At the centre of Earth, there will be a very tiny net pull towards all the parts of that ring – the overall effect of which will be an extremely tiny net pull roughly towards the South Pole.

Incidentally, if you could slide down a frictionless tube through the centre to your antipodal (opposite) point on Earth's surface and back, the round-trip would take you 90 minutes – exactly the same time as it takes to go round Earth in a low orbit.

Robin Clegg
Swindon, Wiltshire, UK

The middle man

Hernando 'Stout' Cortés is said to have climbed a tree on the Isthmus of Panama and seen both the Atlantic and the Pacific oceans from his vantage point. Are there actually any locations on the isthmus (or elsewhere) where this is possible?

John Humbatch
NATO HQ, UK

One can indeed see both the Atlantic and Pacific oceans on a clear day from the Republic of Panama.

My grandfather, the Reverend S. Moss Loveridge, lived in Panama between 1900 and 1919 and he records in his autobiography, *Panama Padre*, how he climbed Balboa Hill, which is now known as Cerro Grande and is situated close to the Panama Canal and 5 kilometres from the town of Gorgona. The summit of the hill lies at 300 metres above sea level and from here he was able to see both the Pacific and Atlantic oceans.

Photographs of the excellent views from this vantage point can be found in *A Trip – Panama Canal*, published by Avery and Garrison of Panama City in 1911.

John Loveridge
Brighton, Sussex, UK

John Keats, who was the author of the sonnet 'On First Looking into Chapman's Homer', was wrong when he stated that Cortés stood silent upon a peak in Darién looking at the Pacific Ocean. In fact, it was Vasco Núñez de Balboa (1475–1517) who first did this and after whom Balboa Hill (now Cerro Grande) was named.

In 1511 Balboa joined a Spanish expedition to Darién and he eventually became its leader. He founded a colony there before marching across the Isthmus of Panama and climbing a mountain from where he sighted the Pacific Ocean. He was the first European to do this, and subsequently took possession of it for Spain.

Brian Wimborne
Isaacs, ACT, Australia

The Isthmus of Panama is, at its narrowest point, 61 kilometres wide. From the middle of the isthmus, the horizon would have to be at least 30.5 kilometres away for both the Atlantic and Pacific oceans to be in view from a single vantage point.

The necessary height of the vantage point can be calculated from trigonometry, by mentally constructing a right-angled triangle, with a hypotenuse that is 30.5 kilometres long, atop a sphere of the Earth's diameter. I recommend the lazy man's way around this, which is simply to look it up in a nautical almanac.

According to *The Amateur Pilot*, by J. E. Milligan, published by Cornell Maritime Press, a horizon 30.5 kilometres away can be seen from any vantage point which exceeds 85 metres above sea level.

The Times Atlas of the World tells us that near the centre of the Isthmus of Panama there are several geographical features which exceed 85 metres above sea level. The prime candidate is a ridge in the Cordillera mountains running roughly east to west and which is parallel to the Chagres river.

The 290-metre peak on this ridge is almost exactly at the halfway point on the old Spanish mule trail from the galleon port of Nombre de Dios on the east coast to old Panama City on the Pacific.

From this central peak the pilot book tables tell us that an observer would have a horizon 65.5 kilometres in every direction. This is quite far enough to see old Panama City 26 kilometres to the southwest and the Atlantic coast 35.5 kilometres to the northwest, and also some 32 kilometres out to sea in both directions.

There is a clear historical record that the earliest explorers of Panama could and did see the Atlantic and the Pacific from the same vantage point. In *Sir Francis Drake Revived*, the 1592 account of Drake's raid on the Panama treasure trains, the text reads:

> … the chiefest of the Cimaroons [escaped slaves] took our Captain [Drake] by the hand and prayed him to follow him if he was desirous to see at once the two seas … Here was a goodly and great high tree, in which they had cut and made divers steps to ascend up near unto the top, where they had also made a convenient bower, wherein ten or twelve might easily sit: and from thence we might without any difficulty plainly see the Atlantic Ocean, whence we came, and the South [Pacific Ocean] …

Dale McIntyre
Dahran, Saudi Arabia

There is a place in Costa Rica where you can see both the Atlantic and the Pacific oceans. This is on top of Volcán Irazu, a volcano near the city of Cartago. It is 3434 metres high and you can go safely to the top because the vent has been dormant since a huge eruption in 1965.

To see both oceans it is best to go in the dry season, which runs from December until April, and as early in the morning as you can get out of bed, because later on in the day, as the heat increases, the air tends to get quite hazy.

Huup Dassen
Geldermalsen, The Netherlands

You can see both oceans from Costa Rica's highest peak, the 3820-metre-high Mount Chirripó. I took the beautiful and very popular trail that leads to the summit with my brother in 1992.

The Chirripó is the highest peak of the Cordillera de

Talamanca (a part of the Cordillera mountains) and is therefore a perfect viewpoint situated almost midway between the oceans, each lying about 120 kilometres away.

Ralf Ludwig
Kiel, Germany

Outside Central America, there are other places where you can see both the Pacific and Atlantic oceans at the same time. One is, of course, Cape Horn at the southernmost tip of Chile. And the tip of the Trinity Peninsula on the continent of Antarctica should be fine for viewing both oceans too.

Obviously you can see them both from space. I'm sure NATO has the resources. I suggest the questioner asks his superiors nicely.

Rick McCallister
Columbus, Mississippi, US

Back and forth

Tides are caused by the moon and the sun and contain a lot of energy, but where does it all come from?

P. Perkin
By email, no address supplied

There is a potential misunderstanding here. The rise and fall of water masses during tides uses up comparatively little energy because in reality little movement happens. When viewed on a global scale, the water masses forming tidal bulges on both sides of Earth (facing and away from the moon) are not moving with respect to the moon. It is the rotation of Earth beneath them that changes.

Mainly due to the presence of land masses on Earth,

this rotation beneath the water does not occur completely smoothly, and energy is indeed dispersed in the tides for the generation of eddy systems, beaches and river mouths. The energy of these tide-related processes is supplied by the rotational energy of Earth, a process known as tidal braking or tidal friction.

Tidal braking causes Earth's rotation to slowly decrease. This decrease in the rotation rate increases the length of a day by approximately 2 milliseconds every century. Although this lengthening of a day is imperceptibly small for most practical purposes, the mismatch between the length of a day now and a day 100 years ago is sufficiently large to bring the time recorded by atomic clocks out of step with astronomical time. Every few years it is necessary to introduce leap seconds to synchronise them.

Frederick Tilman
Cambridge University, UK

The power the tides tap is the rotational energy of Earth, relative to the moon and sun. The moon orbits Earth once a month, and Earth spins on its axis once a day. The effect of the tides is like a brake, causing the spin of Earth to slow while, to conserve angular momentum, accelerating the moon to a higher orbit.

As a result, the length of a day is slowly increasing. This has been measured by examining eclipse records dating back to 600 BC, because the position at which an eclipse can be observed is a sensitive indicator of how far Earth has turned between that time and today. Laser ranging to the moon has also shown that the moon is moving away at about 4 centimetres per year.

Christopher Hughes
Bidston Observatory
Birkenhead, Merseyside, UK

Interestingly, in planetary systems where the moon orbits in the opposite direction to the planetary spin, the effect of tides is to slow the moon down and its orbit lowers, presumably until it is dragged into the planet. At least Earth's future generations don't have that to worry about.

Simon Austin
Shepshed, Leicestershire, UK

Against the grain

Why do sand particles on a beach or dunes seem to reach a certain grain size and then reduce no further? After millions of years, shouldn't most sand have become dust?

Keith Minto
Holt, ACT, Australia

The grains that we see in desert sand dunes have been deposited mainly by wind action. These will generally have originated in other parts of the desert where there are bare rock surfaces that are constantly being weathered by exposure to sun, wind and water – the last of which is a surprisingly powerful weathering agent in deserts.

The result is a build-up of fragments of various sizes: boulders, pebbles, sand grains and dust. These last two, being smaller, can be removed by the wind and transported hundreds of kilometres, either in suspension high in the atmosphere, or by saltation – the process of bouncing along the ground.

The maximum grain size that can be transported by the wind is proportional to the wind speed – faster winds will move larger particles. This means that the larger particles are deposited when and where the wind speed drops, which is often in low or flat terrain. So sand grains of around a certain size accumulate in great masses in lowland basins, while the smaller fragments can be carried further; dust from the Sahara quite frequently falls on the UK, for example. The result is that dunes are made up of grains mainly of the same size.

Similar principles apply on beaches, although the movement of particles is also affected by a variety of additional processes such as wave action, tides, offshore currents and long-shore drift – sand creep caused by waves approaching the beach obliquely. How effective each mechanism is at moving particles depends on its energy, so each will deposit particles in a different location. For example, wave action can sort beach material so that shingle will accumulate as a ridge high up the beach, while sand will only be exposed at low tide. Or long-shore drift may carry sand to one end of a beach, leaving shingle at the other.

Of course, all these fragments – boulders, pebbles and sand – may gradually be broken down into finer particles, so that we might suppose all the world's rocks should by now have been reduced to a mass of dust blanketing the continents. But this does not happen because deposits of sand and dust gradually get compressed and cemented together to form new rock – the sandstones and mudstones. Nor does the planet run out of sand and dust, because bare rock surfaces are constantly exposed to weathering processes, and there will always be new rock exposed as a result of tectonic movement.

Michael Ghirelli
Hillesden, Buckinghamshire, UK

If the sand in a coastal system is too fine relative to the energy of the waves then it will stay in suspension in the water and will not be deposited. So for a beach of dust to exist, the

environment would have to be profoundly calm, and the dust-like sand would have to be kept wet in order to prevent the wind from claiming it. Most beaches are not like this.

Dunes are deposits of wind-blown sand, and for the sand to be deposited the size of the grains must exceed the carrying capacity of the wind. Sand dunes are innately dry places and there is no way that dust-sized particles could hope to stay put in these areas, however weak the wind may be.

Desert dunes exist in gigantic systems, whereas beach dunes form only a narrow band running along the back of some sea beaches, and are created by gusts from the sea that transport sand up from the beach. Yet both systems result from the same key process of wind-borne matter being deposited when the wind becomes too weak to keep it aloft. Of course, even the tiniest sand grains will be deposited somewhere, but they will be highly dispersed and will not form dunes.

Edward Davies
Fareham, Hampshire, UK

Inside out

What would happen to the world as we know it if Earth were hollow below the crust, assuming it didn't collapse inwards?

Terrence Douglas
Halifax, Nova Scotia, Canada

If Earth were hollow we would be in danger of death by suffocation, thirst, frying, starving, freezing and drowning, in that order.

A hollow Earth would not have enough mass to hold on to an atmosphere by gravity, and all the surface water would boil away. If the crust had enough mass to make up for the hollow centre, there would be no magnetic field, which is generated by Earth's liquid iron interior. Compasses wouldn't work, and some migratory animal species might get lost, but those would be the least of our worries because deadly radiation from the sun and outer space could then penetrate to Earth's surface.

If this could be solved, then presuming we could grow gills we could live underwater. We'd need to, because within a million years the continents would have eroded to little more than sandbanks, and the sea level would rise because of all the sediment dumped in the oceans by the rivers. It is only subduction of tectonic plates – where one plate moves under another – and mountain building, created by the same convection currents in the interior that create Earth's magnetic field, that keep uplifting the land to compensate for erosion.

Volcanic eruptions and subduction also play an important role in regulating the carbon dioxide levels in the atmosphere. On a hollow Earth without these processes, plant growth could cease entirely because of all the carbon reaching the ocean floor through erosion, and Earth would enter a period of deep freeze, deprived of the essential warming effect – and food supply – that carbon dioxide gives us now.

Hillary Shaw
School of Geography
University of Leeds, UK

A hollow Earth would weigh only about 5×10^{22} kilograms, roughly a hundredth of the present Earth.

The escape velocity for this new world would be about 40 centimetres per second, so the force required to jump to a height of just 1 centimetre on our normal Earth would send you hurtling into space. The same applies to our atmosphere: virtually all of it would have been lost as the gas particles acquired a velocity far greater than the escape velocity.

It would be a pretty grim world. There would be no spectacular volcanoes, and no source of lava. However, one could save time and money on flights from the UK to Australia by cutting holes on opposite sides of the globe and hopping through one of them – wearing protective clothing of course. The journey time would be a mere 7.5 hours.

Stephen Patrick Elliott
University of Cambridge, UK

A hollow Earth would have been highly susceptible to destruction from the meteorites which bombarded Earth in its early history. Even assuming it was not destroyed by the battering, it would be a barren place.

Without the heat, convection processes, and source material of the core and the mantle there would be no tectonics and volcanism, therefore no atmosphere, no life and probably no water.

Laura Davies
Earth Sciences Department
University of Bristol, UK

Spinning onwards

Erosion redistributes mass from the mountain heights to the bottom of the sea. This must reduce Earth's moment of inertia and so increase the planet's rotational speed and decrease the length of the average day. On the other hand, humans dig holes, and build skyscrapers, shifting the mass-balance in the opposite direction. Are any of these effects of any significance?

Tim Threlfall
Shenton Park, Western Australia

Earth's moment of inertia, which is approximately 10^{38} kilogram-metres-squared, is pretty big compared to any effects we humans have on the system. For example, at the Bingham Canyon copper mine in Utah, something like 4 cubic kilometres of material has been excavated and dumped on the surface. This will have increased our moment of inertia by a figure so small that Earth will slow by an amount which will add up to only 0.03 seconds in a million years.

Volcanic eruptions and earth slides cause material to be lowered, so speeding up the planet's rotation, but only by a similar amount. Construction and demolition of buildings have even less effect. The most significant human effect would appear to be burning fossil fuels, which takes carbon from the crust and pumps it up into the atmosphere, so increasing the planet's moment of inertia, which increases the day length. At current consumption rates, the planet is slowed by the equivalent of 0.3 seconds every million years. This could be compensated for by getting everybody on the planet to run west as fast as they could. However, we would all have to keep it up for a million years.

Terence Collins
Harrogate, North Yorkshire, UK

The biggest movement of mass towards or away from the centre of Earth is accounted for by evaporation and precipitation which, in the short term, cancel each other out. On average, each square metre of Earth gets 1000 millimetres of precipitation a year, or 1 million tonnes per square kilometre. In the longer term, climate changes may cause ice mass to accumulate at the poles, which are close to Earth's axis of rotation but may possibly still affect the speed and axis of rotation a little.

Erosion rates from the land, which covers only 29 per cent of Earth's surface, average around 100 tonnes per square

kilometre per year, though some mountainous areas average erosion rates of more than 1000 tonnes. Not all this sediment reaches the sea – some ends up in lakes, reservoirs or inland deltas. Fortunately for land animals like ourselves, the elastic nature of Earth's crust and the subduction processes, where ocean plates move under continents, mean the land rises in compensation as it is planed down, and sea sediment is ultimately recycled up in volcanic mountains and uplift zones such as the Himalayas. Without these, in about a million years all the land would have eroded into rising oceans, leaving only a few sandbanks above sea level.

Ice accumulations may cause movements of Earth's crust, as may volcanic eruptions, by releasing dust and CO_2. These cause long-term movements of mass both away from and towards the centre of Earth. However, given Earth's radius of 6400 kilometres, a shift of crust upwards even to the height of the Himalayas – less than 10 kilometres – is not large.

Compared with these figures, the construction efforts of a few billion humans are puny indeed. Anyway, when a building is erected, mass may well move closer to Earth's centre. Building causes erosion and, if the gravel and sand for the concrete is quarried in a location a few metres higher than the building site, net mass also moves towards the centre of Earth. Buildings in general are eventually demolished and may be used for landfill, which may also move material downwards. Tunnelling and mining construction will cause material to shift upwards, but as abandoned mines fill with water and subside, material shifts down. It is likely that the construction and demolition efforts of *Homo sapiens* worldwide on any given day will largely cancel each other out.

Hillary Shaw
School of Geography
University of Leeds, UK

High and dry

I accept that it is unlikely, but is it at least theoretically possible that a fault in Earth's crust on an ocean floor could drain an enormous quantity of seawater? What would be the effect, not only on life on Earth, but on the planet's structural integrity, if sea levels underwent a colossal drop? Of course it's hypothetical, but indulge me.

Paul Barrett
Radstock, Somerset, UK

This is theoretically possible only in the sense that nothing is ever absolutely impossible.

Your hypothesis requires the improbable assumption of an enormous void, or a series of interconnected voids, under the ocean floor, containing nothing more than a gas. If the overlying rock were breached by a fault, any gas would either bubble to the surface or be compressed to a much smaller volume by the weight of the water above, lowering sea level as water drains in.

The effect on the planet would depend on the circumstances. If the void were close to the surface, where the surrounding rock is cool, the planet's structural integrity might be improved. A void filled with an incompressible liquid would be more stable than one containing only a compressible gas.

If the void were a little deeper, the resulting geyser might be a spectacular tourist attraction for anyone left to see it. Still deeper and the intense heat would suddenly generate huge volumes of steam and could blow away large portions of the planet, possibly all of it. After all, we are considering the improbable.

The lowered sea level alone would cause dramatic climate changes. If enough water were drained away to significantly reduce the surface area of the oceans, the resulting decrease in evaporation and rainfall would be catastrophic. Life as we know it would have to retreat to the few remaining niches, which would probably not support our species and certainly not our civilisation. For us, evolution would have to start again. Eventually intelligent life might emerge, this time perhaps with better things to do than ponder such bizarre scenarios.

Peter Bauer
Tuckerton, New Jersey, US

The oceans fall and rise by hundreds of metres every time an ice age starts and ends. The process takes centuries or millennia, and each time life replaces life.

Beneath the sea some tectonic plates continuously slide under others, taking many cubic kilometres of seawater with them. But it takes more water than that to affect ocean depths noticeably, and in any case that water mostly re-emerges in volcanic eruptions. This explains why volcanoes near subduction zones emit large quantities of water.

As for sea-floor chasms opening catastrophically, this is highly improbable. For a chasm to open, something must make room for it. That would require either an abrupt swelling of Earth's interior, or a sudden piling up of sea-floor plates. Either process would require such unthinkable energies that sea level would be the least of our worries. In practice, sea floors continually split open at mid-ocean ridges, and magma from beneath immediately congeals and fills the cracks.

Jon Richfield
Somerset West, South Africa

I once attended a lecture by the famous Colonel Brian D. Shaw, a British army explosives expert, in which he demonstrated the explosive power of 1 millilitre of water when it was instantaneously vaporised. He did this by suspending sealed ampoules of water above Bunsen burner flames.

After about 15 minutes, during which he continued his lecture, there were almighty explosions which toppled his safety screens and made the audience jump out of their seats.

He pointed out that these were the results from only a tiny volume of water, and that when Krakatoa erupted it involved the instantaneous vaporisation of 2 cubic kilometres of water. Two-thirds of the island disappeared. I imagine therefore that if more water than this fell through Earth's crust into the molten interior, we might be less concerned with the drop in sea level than with the resulting explosion.

Kenny Walker
Dunblane, Perthshire, UK

Rise and fall

Why are the largest tide ranges in the world – of up to 16 metres – found in the Bay of Fundy, on Canada's Atlantic coast?

Peter Buckley
Toronto, Canada

To understand what happens in the Bay of Fundy, start with a hand basin half-full of water. Push down on the surface on one side with the palm of your hand and the water will rise on the other, after which it will slosh back and forth like a liquid see-saw. By pushing down even more on each side at the same time as the level on that side is falling, the rise and fall of the surface will increase – and can be made to overflow the rim of the basin. The sloshing of the water has a natural frequency and your additional input resonates with it, increasing the amplitude of the see-saw wave. At the central

axis of the basin the level remains unchanged, although water moves to and fro horizontally.

Now imagine the basin cut vertically down that central axis and consider just half of it. The half that is left corresponds to the Bay of Fundy, the axis-edge marks the opening of the bay, and the missing half of the basin is replaced by the open Atlantic Ocean. An incoming tide effectively appears in the ocean as a huge wave advancing towards the bay. As it reaches the continental shelf at the bay's opening it plays the role of the high half of the see-saw and happens to coincide with the low water level at the far end of the bay. By the time this wave has moved to the innermost section of the bay – raising its water level to a peak – the dip in the ocean surface corresponding to the low tide has reached the continental shelf.

The exceptionally high tides occur because the successive incoming tides appear at nearly the same frequency as water sloshing into and out of the bay, just as happened in the hand basin. It is a resonance effect.

If the Bay of Fundy opened directly into the Atlantic it would not have such high tides because the natural period of water moving to and fro in it would be only about 9 hours, which is not close enough to the 12.5-hour period of the tides to lead to significant amplification of the wave motion. However, the bay opens into the Gulf of Maine and together they have a natural frequency of just over 13 hours.

Richard Holroyd
Cambridge, UK

Answer in depth

When will Challenger Deep cease to be the deepest ocean trench? And how will this occur?

Zak Harris
Birmingham, UK

The Mariana trench, whose deepest point below sea level is Challenger Deep (10,898 metres), is the result of the planet's largest tectonic plate, the Pacific plate, slipping under its neighbour, the Mariana plate. This process of one plate sliding under another is called subduction.

Ocean trenches are formed over millions of years, and the simple answer to the above question is: nobody knows. Challenger Deep may always be the deepest point beneath sea level because of the relative masses of the plates that form it.

Interestingly, Challenger Deep is not the point on the seabed that is nearest to the centre of the planet. Because Earth is flattened at the poles, that point on the seabed is likely to be in the Arctic Ocean.

Tony Holkham
Boncath, Pembrokeshire, UK

Exactly when the forces that created Challenger Deep will moderate we cannot say. What is clear is that as soon as they stop, the hole will begin to fill with biological detritus, chemical concretions and the products resulting from erosion of the surrounding topography. Once that begins, the process should be pretty quick and it might be difficult to locate the trench 1 million years or so after its generating forces have abated. Structures such as this tend to be geologically short-lived. By way of comparison, a 1-million-year-old lake is a pretty old one – it will usually have silted up, dried up or eroded long before that.

The reason the African Great Lakes are still there is that the Great Rift Valley is still rifting – they are, in fact, an ocean in embryo. It is a similar story with oceanic deeps; remove the forces that cause them, and they pass away like dimples in foam.

Jon Richfield
Somerset West, South Africa

Reach great heights

When will Mount Everest cease to be the tallest mountain on the planet? And how will this occur?

Jade Harris
Birmingham, UK

This is a good question, but the short answer is that nobody knows. Let's clarify some terms first: Mount Everest is not the tallest mountain on the planet by a long way – that accolade goes to Mauna Kea in Hawaii, although much of it is below the ocean. The summit of Everest is not even the furthest point from the centre of the planet – that is believed to be the summit of Chimborazo in Ecuador. The summit of Everest is, though, the highest point on the planet above mean sea level.

It seems to be generally accepted that the Himalayas, in which Everest is located, are still rising by about 5 millimetres per year as a result of the movement of tectonic plates in the region. Geological time is measured in millions of years, so the summit of Everest is likely to remain the highest point on the planet for many years yet.

Other relatively new ranges are rising too, but, because the Himalayas include most of the top 100 highest points on the planet, it's unlikely that Everest – more than 200 metres taller than its nearest rival – will be overtaken for some time. The best answer we can give, I guess, is yonks.

Tony Holkham
Boncath, Pembrokeshire, UK

Hot youth

All of the radioactive elements that made up the early Earth are apparently the hot ash of ancient supernova explosions. This means we've been working through their half-lives for at least 5 billion years. How much hotter than today was the interior of our planet when it first came into being? What would such heat mean for tectonic activity and the evolution of life in our world's feverish youth?

Pat Sheil
Sydney, Australia

It would take a major study to calculate details of the isotopic make-up of the planet 5 billion years ago. I suspect that internal temperature would have been dictated by gravity making the planet contract and impactors hitting its surface, rather than isotopic decay or nuclear fission.

Our present concerns with nuclear decay vanish over such timescales. Consider that plutonium-242 has a half-life of more than 370,000 years. There are several thousand half-lives in a billion years, so even if our galaxy were made of solid plutonium-242 at the start, none would now remain. With a half-life of over 710 million years even uranium-235 would by now be reduced to less than 1 per cent of the original. Strontium-90 would vanish in less than 1 million years and would be gone before the planet had clumped together from the remnants of supernovae.

In short, we need not consider isotopes with lives much

shorter than that of uranium-235. Having said that, self-sustaining nuclear chain reactions are possible. Natural nuclear fission reactions took place around 1.7 billion years ago in uranium-235 deposits in Oklo, Gabon. They ran for a few hundred thousand years, with an average power output of 100 kilowatts during that time.

We don't know exactly when life started, but it was a good 500,000 years after Earth formed, when tectonic activity had calmed to something less than the free convection of molten rock.

Antony David
London, UK

If nuclear decay didn't heat the planet, does anyone know what did? – Ed.

The initial heat of Earth was caused by the collisions of smaller objects that came together to make it. Gravitational attraction accelerated objects towards each other, and when they struck their kinetic energy converted to heat.

As time went on, the amount of energy contributed by objects falling onto the proto-Earth became larger as the planet and its gravity became greater. Towards the end of the process, any object hitting Earth would contribute at least 60 kilojoules for every gram of its mass. If that heat were confined to the colliding object, it would heat it to thousands of degrees. This heat did not have time to radiate away as the planet grew, so the result was a hot Earth.

Besides this, Earth produced heat through radioactive decay. Early on, the amount of heating per unit volume was greater than at present. Today, heating is almost all due to the decay of uranium-238, thorium, and potassium-40, in roughly equal measure. But 4.5 billion years ago, the amount of uranium-235 was close to the amount of uranium-238 we have today (instead of being a minor component) and uranium-235 produces heat much faster than uranium-238. There was also about 13 times as much potassium-40 as there is today. The presence of these nuclides helped to heat up the interior of Earth during the first billion years of its existence.

Eric Kvaalen
Les Essarts-le-Roi, France

An earlier answer to this question said Earth's internal heat is generated by the radioactive decay of uranium, thorium and potassium. There was, however, no mention of the gravitational heating caused by the pull of the moon in its elliptical orbit around Earth. That pull must have been considerable in the distant past, when the moon was much closer in to our planet. The heating effect is particularly evident today on Jupiter's satellite Io.

In Earth's case, has anyone calculated how much heating is due to radioactive decay and how much to the moon's gravitational pull, both now and in the past?

Neil Macnaughtan
Edinburgh, UK

Can't stand the heat?

If heat destroys magnetism, why doesn't the heat inside Earth destroy its magnetic field?

Martin McCann
Liverpool, UK

Magnetic fields are generated by the movement of electrical charges in a process called electromagnetic induction. A single electron orbiting the nucleus of an atom generates only a weak magnetic dipole. For a piece of solid material to be a permanent magnet on a macro-scale, the dipoles need to align, which is what happens in materials like iron and nickel. However, when the material is heated above a certain temperature, the Curie temperature, thermal vibration of the atoms starts to push the dipoles out of alignment, and the material loses its bulk magnetism. So heat does not destroy magnetism; it merely disrupts the alignment of the dipoles such that, at a macro-scale, there is no longer any net magnetic field.

Earth's magnetic field is believed to be generated by electric currents in its molten core, which is mainly iron. These electric currents are formed by convection currents of radioactive heat escaping from the core. If the core was not hot enough to be liquid, no flow would occur and no magnetic field would be induced. A similar process occurs in the sun to produce its magnetic field.

Simon Iveson
Chemical Engineering Discipline
University of Newcastle
New South Wales, Australia

2 Space

Light star

If the sun is a star just like all the others, why does it appear yellow rather than white?

John Berry
Newark, Nottinghamshire, UK

As a beam of light passes through the air, blue light tends to be scattered while the red and yellow continue to pass through. This is why the sun looks red at sunset as the light passes through a great thickness of atmosphere, and also why the sky is blue.

If you put the beam back together, adding the light from the blue sky and the yellowish sun, it would appear white. This addition of light covering the full visible spectrum is what happens when you look at a field of snow.

The same thing happens to starlight. But the cells in your eye need a large input of energy to detect colour, so dim stars seem colourless. If the sun were a distant star it too would look white.

Spencer Weart
Center for History of Physics
Maryland, US

Contrary to popular belief, stars come in many different colours. This is because their colour is intrinsically linked to their temperature.

Young, hot stars will be white, while cooler, older stars will be red. Our sun is a medium-sized star, burning at a medium rate. This is reflected in its yellow colour.

Sion Amlyn
Trefor, Gwynedd, UK

There are two main reasons why stars appear white, even though they really cover a wide range of colours from deep red (cool stars, less than 3000 °C) to bluish-violet (hot stars, greater than 30,000 °C).

First, the human eye is poor at detecting colours at low light levels. For example, all cats look grey in moonlight. Hence, most stars, especially the fainter ones, appear white.

Secondly, many of the stars visible to the naked eye are genuinely white or bluish-white, being among the hottest and most luminous ones in our region of the galaxy, such as Rigel in Orion.

Notable exceptions are Betelgeuse in Orion and Antares in Scorpius, which are highly luminous, cool red giant stars near the end of their lives, although these do appear distinctly red when compared to other stars.

Other stars of the same temperature (about 5500 °C) and size as the sun, or even cooler and smaller ones, are either too dim for their colour to be distinguished by the naked eye or cannot be seen at all without a telescope.

Paul Hatherly
University of Reading, Berkshire, UK

Star quality

Stars appear as dots of bright light in the night sky. As with our sun, every star represents an immense light source that is dazzling close up. So given the brightness of stars, why are the intervening spaces, when viewed from Earth, black?

Graham Mays
By email, no address supplied

Stars are so far away that the amount of light reaching us is small. A relatively radiant star, such as Antares, is as bright as a single candle a kilometre away. This is not enough to dazzle you. If your eyes are able to focus at infinity, the light of a star falls on just one rod or cone of your retina (or a few if there is dispersion). If your eyes are out of focus, each star will appear as a little disc, but dim stars will not really be visible, so you will still see black between the bright ones.

Eric Kvaalen
Les Essarts-le-Roi, France

The naked eye can see perhaps 9000 stars. Using medium-power telescopes this number grows to millions, and modern observatories may detect billions. Estimates vary as to the number of stars in the universe – or the observable universe, because what we see is limited by the speed of light. A minimum figure commonly cited is 10^{22}.

Why, then, are we not ablaze at night? This is what is known as Olbers' paradox, named after the 19th-century German astronomer Heinrich Olbers. At the time, the universe was thought to be static and set in an infinite sea of stars. Now we have considerably more knowledge about our cosmos.

The simplest, most widely accepted hypothesis is that many of these stars are so far away that their light has not

yet reached us, and perhaps never will – at least not in a form we can see. The expanding universe delays or prevents light reaching us because of the distances involved, and maybe because galaxies recede from us at speeds effectively greater than that of light. And the wavelength of the light radiation lengthens as galaxies recede from our standpoint. This is known as redshift because the receding light source is shifted to the red end of the light spectrum.

Interestingly, the essence of this explanation is the same as that given by Edgar Allen Poe in *Eureka*.

K. D. Perring
Ashford, Kent, UK

Thanks to all who sent in the following section from Poe's essay, Eureka, *written in 1848 – Ed.*

> Were the succession of stars endless, then the background of the sky would present us an uniform luminosity, like that displayed by the Galaxy – since there could be absolutely no point, in all that background, at which would not exist a star. The only mode, therefore, in which, under such a state of affairs, we could comprehend the voids that our telescopes find in innumerable directions, would be by supposing the distance of the invisible background so immense that no ray from it has yet been able to reach us at all.

Pulling power

To escape Earth's gravitational field, a spacecraft has to travel at escape velocity. Why can't it travel at lower speeds? If a rocket-powered device has enough thrust to lift its weight, surely it will eventually make it to space?

Graham Drake
Leighton Buzzard, Bedfordshire, UK

The concept of escape velocity applies only to unpowered projectiles, not powered rockets. Unfortunately, the definitions of escape velocity given in many textbooks do not make this clear. Obviously, a rocket rising vertically at a low speed would eventually reach space if it kept going at the same speed. But in practice, rockets run out of fuel and become unpowered projectiles. If they have not reached escape velocity when this happens, they will not escape Earth's gravity.

Peter Lafferty
Wadhurst, East Sussex, UK

A spacecraft needs energy, not velocity, to leave Earth's gravitational field. You can leave Earth as slowly as you like, providing you do enough work against gravity along the way.

Suppose a spacecraft has a weight W on the launch pad. If the thrust from the rockets is greater than this, the spacecraft will move upwards. As it does so, it gains gravitational energy. If that energy gain is at least WR (where R is the radius of Earth), gravity can never pull that spacecraft back to the surface.

The least efficient way of launching a spacecraft is to have the thrust slightly greater than the weight. The spacecraft moves up very slowly and runs out of fuel before it gets very far.

The most efficient strategy is to launch the spacecraft very quickly, giving it kinetic energy. It can then be left to coast the rest of the way, slowing down as it gains gravitational energy. The spacecraft will have enough kinetic energy if its launch velocity is at least 11 kilometres per second.

However, the high acceleration involved in such a rapid launch, to say nothing of the heat generated by friction with the atmosphere, means that a compromise strategy is needed. A constant acceleration of 30 metres per second gets the spacecraft out of Earth's atmosphere in about 30 seconds with a speed of about 1 kilometre per second. About 5 minutes after launch the spacecraft will have travelled 1500 kilometres and will reach escape velocity. Earth's gravity will have dropped only by about a third. Nevertheless, the rockets can now be turned off.

Michael Brimicombe
Aylesbury, Buckinghamshire, UK

Moonflight

If I were an Olympic high jumper would I be able to jump high enough to escape the moon's gravity? I suspect not, but some astronauts on the moon's surface seem to float in the air for a long time. Exactly how high and how fast would I have to jump to escape the moon's pull, or simply to fly over its surface? As there is no atmosphere, are my aerodynamics irrelevant?

Sandie Mountford-Jones
Stoke-on-Trent, Staffordshire, UK

Sadly, even an Olympic high jumper would be unable to break free of the moon's gravity. However, they would be able to leap a lot higher than their earthly counterparts, because the moon's surface gravity is about six times weaker than Earth's.

The current world record for the high jump is 2 metres 45 centimetres, therefore the moon record would be somewhere in the vicinity of 15 metres.

To escape the moon's gravity entirely you would have to jump at what is called the escape velocity. This is the speed you need to reach before you can entirely escape a body's gravitational field at a single kick. For Earth, this is about 11 kilometres per second. To escape the moon's gravity, the speed is more leisurely but still way beyond a high jumper's take-off speed.

Escape velocity is the speed which is needed to escape with no further propulsion, but if you gave a slug a very long ladder and sufficient lettuce, it could keep slithering along all the way to infinity without ever having to travel at escape velocity. And so it was with the Apollo lunar modules, which didn't need to reach escape velocity to get back into orbit because they had an engine providing sustained power.

To fly over the surface of the moon would effectively require going into orbit (or simply leaping a very long distance). An orbit is really just a jump which never comes back down because the body over which you move is spherical and its surface is falling away from you as fast as you are falling towards it. In order to go into orbit just above the surface of the moon, a high jumper would need to leap at about 1 kilometre per second – still rather out of range.

So all the moon is really good for, if you are a high jumper, is setting new extraterrestrial records. You'll need to find a planet significantly smaller before you can fly off into the sunset – even with a high jumper's legs.

Andrew Steele
Newport, Shropshire, UK

To escape the clutches of the moon's gravity you would need an escape velocity of 2.38 kilometres per second, about a fifth of the 11.2 kilometres per second required to escape from Earth. Gravity is six times weaker on the surface of the moon than it is on Earth, so in theory athletes on the moon would be able to jump six times as high as they can on Earth. The moon record for the high jump would then be 14 metres 70 centimetres. Pole-vaulters would be able to clear the twin towers of the old Wembley Stadium in London and TV programmes could dispense with slow-motion action replays because the jumps would take six times longer. However, all this could be achieved only if athletes could run as fast on the moon as they can here on Earth and, given the difficulties experienced by moonwalking astronauts, this seems unlikely.

Nonetheless, those who can clock a respectable 11.4 seconds for the 100 metres on Earth would be running fast enough to launch themselves from the asteroid Toro: discovered in 1964, Toro has a radius of only about 5 kilometres. Those who are not quite as nimble on their feet might want to choose one of several smaller asteroids. With a radius of only 1 kilometre, a brisk walk on Geographus, for example, might do the trick. Its gravitational field strength is less than 1/6000th that of the moon.

Assuming, however, that you are restricted to the moon but really needed to escape its gravitational field, you could climb a ladder to escape. According to the UK government's Department of Health, the average man needs to consume about 2550 calories per day. Our bodies are about 40 per cent efficient at converting food into useful energy. Assuming no other losses and that the food consumed was used solely for climbing the ladder, it would take an average man 49 days of continuous climbing, 24 hours a day, to escape the moon's gravitational field.

Mike Follows
Willenhall, West Midlands, UK

You would not be able to jump very high on the moon – the

slow movements of astronauts are only relative to our perceptions, which have evolved on Earth. Even a rifle bullet would not escape from the moon if fired vertically upwards.

You would need a very small body such as the asteroid Eros, which passes relatively close to Earth, to have a chance. Eros is 34 kilometres long and has an estimated escape velocity of 3.1 metres per second at the pole – about the speed of an Olympic high jumper.

However, you can also leave the surface travelling horizontally if you achieve orbital speed. This speed is equal to the escape velocity of the asteroid multiplied by 0.707 (for a sphere) minus any speed due to the body's rotation. On the moon you would need to be travelling at 6000 kilometres per hour, but you could orbit Eros at just 6.5 kilometres per hour, which would give a new meaning to the term spacewalk.

Jerry Humphreys
Bristol, UK

Even if we allow an athlete the use of a stored-energy device, such as a catapult, the amount of energy required to achieve escape velocity from the moon is equivalent to 17 ascents of Mount Everest, starting from sea level. It would be a long and exhausting process.

However, using stored energy or a system of levers, it might be feasible for a human to put a small object into orbit about the moon. A golf ball can exceed 320 kilometres per hour and levers could transform this into a much higher speed. Additionally, if a weight was swung round on a string while you progressively played it out, because of the lack of air resistance on the moon, the weight could attain the required speed and travel into orbit when released.

Terence Collins
Harrogate, North Yorkshire, UK

The maximum height a high jumper can attain on the moon and the one quoted by most of your correspondents so far is the 'urban myth' height of six times the current high-jump record on Earth.

Actually, the correct figure is about 9 metres, not 15 metres. This is because the energy that a high jumper must expend is that required to lift their centre of mass over the bar, and not that required to lift their feet. The centre of mass of the average human is about 1.1 metres from the ground. Most Olympic high jumpers, who are tall and slim, will have a centre of mass further from the ground than this.

Therefore, the world record holder for the high jump expended the energy necessary to raise their centre of mass by about 1.35 metres, not 2.45 metres (the current record). The position of their centre of mass will be the same on both Earth and the moon and consequently the height jumped on the moon will be six times 1.35 metres, plus the height of their centre of mass above the lunar surface, giving a grand total of 9.2 metres.

So I'm afraid that we are even more puny than originally thought.

Paul Robinson
West Bridgford, Nottinghamshire, UK

Mercury madness

I have asked many people this, including my parents and science teacher, but they don't know the answer. If a thermometer was in space, what would it read?

Joseph Eaton
London, UK

Thanks to those who pointed out the complicating nature of the title. A thermometer using mercury would be useless in space for the following reason – Ed.

A mercury thermometer would stop working at 234 kelvin (−39 °C) – the temperature at which mercury freezes – but there are types of thermometer that would work.

If the thermometer was in Earth's shadow and there was no moonlight, then it would read about 3 K. This would be because of cosmic microwave background radiation at 2.7 K and some infrared radiation from the night side of Earth. If it was in direct sunlight, the temperature would be about 500 K (230 °C), which is why some spacecraft such as the Apollo lunar modules were wrapped in gold foil to stop them overheating.

So the short answer is it would record anything from 2.7 K to millions of degrees, depending on where it is in space.

Andy Biddulph
Burton upon Trent, Staffordshire, UK

In the vacuum of space, there are virtually no particles to conduct or convect heat, so a thermometer would not be affected by these processes. However, it would still be affected by radiation from stars and cosmic microwave background radiation.

Also, the thermometer will radiate away any residual heat until it reaches equilibrium with its surroundings – in this case the same temperature as space. Because collisions with photons emitted from stars are rare, the thermometer will essentially measure only the temperature of the cosmic microwave background radiation, so it would read about 3 kelvin (or −270 °C).

Lewis O'Shaughnessy
Salisbury, Wiltshire, UK

3 Physics

Biggest and best

In simple terms, why does the Large Hadron Collider have to be so physically large, when it is designed to detect particles that are incredibly tiny?

Mike McCullough
London, UK

The Large Hadron Collider (LHC) is a synchrotron, a circular accelerator that uses carefully synchronised electromagnetic fields to accelerate particles to very high speeds. When this involves charged particles on a curved path they release synchrotron radiation, which wastes energy. This is not desirable because most of the particles that physicists are looking for, such as the recently discovered Higgs boson, have large masses and can only be created in high-energy collisions.

The large radius of the LHC's track is big enough to limit the radial acceleration given to the particles, thus minimising the energy that they lose as synchrotron radiation. The superconducting magnets used to control the flow and direction of the particles can accelerate them up to 99.9999991 per cent of the speed of light.

Doris Lee
Burnaby, British Columbia, Canada

The short answer to your question is 'the magnets'.

The size of the LHC is actually a trade-off between three things: the magnets that are available; the energy – or velocity – it is necessary to give the particles; and the feasible dimensions of the structure.

The faster the particles are moving, the more likely you are to see something interesting happen in a collision. So it's important to accelerate the particles, mainly protons, as much as possible.

The protons need to follow a circular path so they can be continuously accelerated by an electric field, and this is done using magnets positioned around the tunnel. The faster the protons travel, the stronger the magnetic fields need to be to keep them on track.

To increase energy there are two possible choices: make the magnets stronger or the accelerator ring larger, so that the particles' path does not need to be bent so much. At some point there is either a technological or financial limit on the strength of the magnets, leaving ring size as the only remaining variable.

However, to keep the costs of the project manageable, the LHC was built in an existing tunnel that housed a previous experiment, called the Large Electron-Positron Collider. So the energy to which protons can be accelerated was actually predetermined by limits of technology and funding.

When people say the next generation of accelerators needs to be the size of the solar system, they should qualify this by admitting that they are assuming existing magnet technology cannot be improved upon.

Michael Luk
By email, no address supplied

It is not just the LHC's circumference that is huge; so are the detectors. It's here that the collisions take place and physicists look for new particles. A detector is packed with all kinds of instruments to measure mass, energy, temperature and so on, and there are additional trackers and triggers right around the beam. In part these explain the detectors' size.

But there is something else. We know that a theoretical particle can be recognised by the way in which it decays into particles we already know exist.

For instance, in the LHC's ATLAS detector, a huge cylindrical device, the Higgs boson will theoretically decay into muons, which are massive relatives of the electron. We can see muons by tracing their paths. But they are very elusive particles and will fly through any substance with enormous speed. So the muon detectors extend from a radius of 4.25 metres from the axis of the proton beam out to 11 metres.

If the detectors were closer to the beam or narrower than this, the muons would fly past before they could be identified. So the size of the LHC has much to do with what we can physically measure with the means available to catch the unseen wonders of nature.

A friend of mine helped build the hardware for the muon detectors in ATLAS, and he and his colleagues kindly took me on a tour of the ring and the detectors when the LHC was almost finished. It was the best trip I have ever been on.

Marieke Nelissen
Amsterdam, The Netherlands

Atomic mushrooms

This summer our high school visited the museum at Hiroshima about the American attack on Japan with an atomic bomb. We also heard stories from survivors and saw pictures of the destruction, some taken from a long way away and showing a giant black cloud over Hiroshima. I wondered why the nuclear bomb always makes a mushroom-shaped cloud?

Hiroko Watanabe
Minato-ku, Tokyo, Japan

When a nuclear device detonates it creates an area of very hot gas and waves of X-ray radiation. This is the destructive component that damages buildings and leads to the appalling death rate. The heat and radiation rapidly heat up the air around them and, as with all hot gases, the air begins to rise, in this case very quickly. This drags up debris, dust and moisture to form the cloud that eventually becomes the mushroom.

At first the mixture of hot air and dust rises vertically, forming the column of the cloud. But as the cloud expands and meets the colder air higher in the atmosphere it slowly cools. Eventually the cloud reaches the temperature of the surrounding air and ceases to rise, but spreads horizontally along air levels that are the same temperature. This spreading leads to the mushroom shape.

While all atomic bombs will produce this effect, hydrogen bombs produce the highest explosion yields with the greatest heat and therefore the most distinctive mushroom shapes. The heated gases rise so high that they pass right through the troposphere and reach the stratosphere. The marked difference in temperature at this point means that the heated gas is incapable of breaking through into the stratosphere and instead spreads rapidly outwards to produce the clearest example of the mushroom shape.

It is not only the atomic or hydrogen bomb that creates the distinctive mushroom-shaped cloud. Any explosion capable of creating the same conditions, such as a massive discharge of conventional explosives, would produce a similar cloud.

Thomas Hewittson
San Diego, California, US

When the hot gas and dust reaches the cool air high in the atmosphere it spreads out, equally in all directions, forming a circular shape. As it cools further it begins to descend at the edges. This leads to the distinctive folding underneath that produces the characteristic mushroom shape.

Nick Pope (and his dad)
Cranfield, Bedfordshire, UK

I have seen a mushroom cloud that was not produced by an atomic explosion. In August 1944, a German V-1 flying bomb, or doodlebug, struck a hydrogen-filled barrage balloon above a Kent barley field. The mixture of explosive, hydrogen and burning cereal produced a perfect example that was perhaps no more than 100 metres high.

Peter Brown
Styal, Cheshire, UK

? Water stones

How do pebbles skim on water? Neither medium seems especially elastic, so how do the stones bounce?

Juan Bandini
Ventimiglia, Italy

For the best results when skimming stones you need a flattish stone. The closer it is to circular, the better. It must be thrown so it is almost horizontal to the water's surface, and also such that its trailing edge hits the water first. It is vital the action of throwing imparts spin to the stone.

Any solid body moving through a liquid experiences forces that oppose its motion. These forces are proportional to the cross-sectional area of the body and the square of its speed. Although only a part of a skipping stone is actually moving through water, with the rest travelling through air, these forces still have an effect on its forward motion.

There is a force exerted by the water at right angles to the spinning stone's surface. It acts at the trailing edge of the spinning stone – because this is where the impact begins – and tends to turn the stone towards the horizontal. Because of its spin, though, the stone behaves like a gyroscope and refuses to change its orientation. Nevertheless, this force reduces the stone's forward velocity somewhat.

There is also a force exerted by the water parallel to the stone's surface, but this force is much smaller and so the stone's velocity is barely changed by it on impact with the water. The net effect of these forces is that the stone flies from the water in a parabolic arc until it hits the water again and the whole process is repeated.

At each impact, the stone loses some of its kinetic energy, which is dissipated in the ripples that are created in the water. And as its velocity is gradually reduced, the impacts become closer together until the energy dissipated is greater than that lost in the impact and the stone sinks.

The minimum initial speed required by a stone varies with its inclination to the horizontal. Experiments show that skimming will not occur if the angle at which the flat surface of the stone hits the water is more than 45 degrees to the horizontal. The slowest speed for skimming is about 2.5 metres per second, when the inclination is about 20 degrees.

The fact that the water itself is not elastic is immaterial, but it is important that it gives way to the stone on impact.

Stones can also be skimmed on wet sand, and even on cloth-covered boards. In such cases, however, there is little or no give in the surface and the frictional grip at impact is sufficient to change the direction of the stone's motion and also cause the stone to overcome the gyroscopic effect.

Skimming stones on water is an age-old pastime. The gunners of naval sailing ships worked out that it could be used to increase the range of cannonballs. These could be made to skip along the surface of the sea and hole enemy vessels near their waterline. However, to do this the cannon itself had to be near the sea surface, which meant the firing vessel needed perilously low ports. Indeed, some ships capsized after taking on water through those open ports. With cannonballs, the spin required by a skimming stone was unnecessary as their spherical symmetry precluded any gyroscopic effects.

Barnes Wallis's famous bouncing bomb, created for the 1943 Dambusters raid on Germany in the Second World War, worked on the same principle. He had to use cylindrical bombs, though, so he figured out he could ensure their stability by giving them spin about a horizontal axis at right angles to their direction of motion on the water.

Richard Holroyd
Cambridge, UK

In a paper published in the journal *Nature* in 2004, the French physicist Lydéric Bocquet and his collaborators revealed some of the secrets of successful stone skimming. They found that the optimum angle of attack is 20 degrees. So, even when the stone is thrown horizontally, the leading edge should be 20 degrees higher than the trailing edge. This maximises the number of jumps by limiting the contact time between the stone and the water, which is proportional to the energy dissipated.

The thrower also imparts spin to the pebble, providing a gyroscopic effect that stabilises its flight and preserves the original angle of attack when it bounces. In the absence of spin, the water would impart a torque on the stone and, because the trailing edge is the first to make contact with the water, this would tend to make it tumble.

The actual physics of stone skimming is not yet perfectly understood. However, the bounce could be understood as a result of the conservation of momentum and Newton's third law: when the stone exerts a force on the water, the water exerts an equal and opposite force on the stone. This lifting force is proportional to the density of the water, the surface area that is wetted and the square of the forward speed of the stone. Also, the bow wave created ahead of the stone when it strikes the liquid might act like a waterski jump, helping to launch the next hop. This minimises the contact time between the stone and the water, which in turn maximises the number of jumps.

Although ensuring the optimal angle of attack as the stone strikes the water, and imparting just enough spin to maintain stable flight are important, there are other factors. Selecting the correct size and shape of stone and having a fast throwing arm are examples.

Given that the urge to skim stones has been with us for thousands of years and the rules – getting the greatest distance or number of bounces – have remained unchanged since the ancient Greeks, perhaps this should become an Olympic sport. In the meantime, the current world record stands at 51 skips, set by Russell Byars in Pennsylvania on 19 July 2007.

Mike Follows
Willenhall, West Midlands, UK

Mix and match

Is it possible for two (or more) ingredients, when mixed, to weigh more than they do separately? If so, what and why? When I make porridge in the morning it certainly seems as if this is the case. If so, have I stumbled on a potential dieting gold mine?

Jamey Barron
Northallerton, North Yorkshire, UK

The simple answer is no. One of the fundamental laws of physics is the conservation of mass and energy. Normal chemical processes do nothing to alter matter, they just rearrange atoms in different ways. Nuclear reactions can convert matter to energy, but as long as this energy is not lost from the system, the system will still weigh the same.

There is certainly no way for a system to gain mass unless there is a flow of energy or matter into it from outside. For instance, a rusting nail weighs more than the original because it has reacted with oxygen from the air.

In the case of porridge, the questioner is seeing a swelling of the oats as they absorb water. The contents of the pot may even swell to exceed their original volume, due to the expansion of gases trapped inside the oat grains during the cooking process. However, the mass of the porridge cannot exceed the original mass of the oats and water before they were mixed together. If anything, there will be a loss of mass due to evaporation of some of the water. Be careful not to confuse volume with mass.

Simon Iveson
Chemical Engineering Discipline
University of Newcastle
New South Wales, Australia

Theoretically, because $E = mc^2$, the mass would increase very slightly if there is an increase in energy. But usually, mixing two substances decreases the energy, otherwise they wouldn't mix – as oil and water do not.

There is one explanation for an apparent increase in the weight, though. When you weigh something, you are actually measuring its true weight minus the buoyancy due to the volume of air it displaces. And weight is different from mass. So if mixing two ingredients results in a smaller volume, then there will be less buoyancy and it will seem heavier. For instance, mixing soda (sodium carbonate) and water will show this effect, but it's a very small one.

I suspect that the enquirer is fooled by the density. The mixture of water and, say, oats is denser than the bulk density before mixing because of all the air that was previously between the dry oats, so it looks heavier even though it's not.

Eric Kvaalen
Les Essarts-le-Roi, France

It is possible for two substances to weigh more after they are mixed than they did separately. Alcohol and water love to mix and form hydrogen bonds. A 50:50 mixture of ethanol and water takes up about 96 per cent of the volume of the separate liquids. Thus a litre of alcohol mixed with a litre of water contracts by about 80 millilitres and thereby displaces 80 millilitres less air than the separate liquids do. And 80 millilitres of air weighs about 0.1 grams. So the mixed liquids are heavier by 0.1 grams because less displaced air means less buoyancy from the weight of displaced air.

David Emanuel
Tulsa, Oklahoma, US

Magnetic marbles?

Is it possible to build a spherical magnet? If we could manage to do it, where would the north and south poles be, and what could such a magnet be used for?

V. Taylor
Milton Keynes, Buckinghamshire, UK

The simple answer to this question is yes. Just grind the corners off a normal bar magnet until it is spherical. The north and south pole field lines would emerge from opposite points on the sphere.

The more intriguing question is whether a monopole can be created – the magnetic equivalent of a particle with either positive or negative electric charge. I'm not aware of a physical law that prevents the existence of monopoles but, despite a long search, I can find no real example of one.

A collection of particles all of the same pole could have many applications. Because they would magnetically repel each other, if the particles were small enough that the repulsion was stronger than their weight, they would behave in a fluid-like manner. Such fluidised particles have many applications in chemical processing because of their heat and mass transfer properties.

Normally, this state is created by passing a fluid up through a bed of particles to keep them in suspension – air is blown up through coal particles to burn them in some power plants, for example. Fluidising particles without a large upwards flow would reduce the energy required and also reduce losses due to attrition because the collisions between particles would be less violent.

This might enable applications that were previously impossible. In particular, fine particles (say less than 30 micrometres) are notoriously hard to fluidise because surface forces tend to make them stick together. However, such particles are very useful because of their large surface-area-to-mass ratio. So, if you find a way to make monopoles, I strongly suggest you patent it.

Simon Iveson
Chemical Engineering Discipline
University of Newcastle
New South Wales, Australia

Your correspondent asks if it is possible to build a spherical magnet. The answer is yes – we all live on one: Earth.

Where would the poles be? The North Pole is where polar bears live and the South Pole is where penguins live.

What's it used for? I suppose the magnetic field's most important role is protecting the planet from the solar wind, which would otherwise strip away our atmosphere. This would mean no *New Scientist*, and nowhere to print this answer.

Perry Bebbington
Kimberley, Nottinghamshire, UK

The first spherical magnets were made in the 16th century or earlier by grinding naturally mined magnets, called lodestones, into spheres. These were called 'terrellas' (little Earths) by William Gilbert, the physician to Queen Elizabeth I.

Gilbert spent much of his life investigating these natural magnets and plotted the first diagram of the magnetic field of a dipole, although he had no concept of field lines. He presciently believed that Earth itself was a giant lodestone with a weathered surface. His book *De Magnete* was published in 1600.

Rod Wilson
Liverpool, UK

Thanks to all who wrote in to tell us that small, magnetic marbles covered in brightly coloured plastic can actually be bought as toys – Ed.

Logs to ashes

I recently attended a bonfire party. There were various kinds of wood being put on the fire, including treated timber, such as old furniture and fences, and recently cut stumps and branches. Nearly all of it burned away, but at the end there was still some ash. Which bits of the wood don't completely burn, and why is there a residue left?

Kevin Slater
Barrow, Cumbria, UK

Like many plants, trees make sugars, cellulose and other organic molecules using carbon from the carbon dioxide in the air. They join it with hydrogen obtained by splitting water from their roots into hydrogen and oxygen. The oxygen is discarded into the atmosphere as part of this system of photosynthesis. All the other elements they need, like nitrogen, phosphorus and metals such as potassium, manganese, iron and zinc, are obtained with the water from the soil.

When wood is burned, oxygen rejoins with the carbon and hydrogen from the organic compounds, releasing stored energy. Oxygen also joins with the trace elements, forming metal oxides and phosphates. It is these compounds that make up the solid ash, which is an excellent fertiliser, giving back nearly all the minerals originally taken from the soil.

Unfortunately, the nitrogen returns to the atmosphere. This explains why slashing and burning trees and other plants to create fields produces good yields for only a year or two before the land becomes nitrogen deficient.

Keith Ross
Villembits, France

There are 17 chemical elements considered to be essential for the growth of most plants. These are carbon, hydrogen, oxygen and nitrogen in large proportions; medium proportions of phosphorus, potassium, calcium, sulphur and magnesium; and tiny amounts of boron, chlorine, copper, iron, manganese, molybdenum, nickel and zinc. These elements are the constituents of the wood that don't burn – the wood ash.

Different plant tissues contain these elements in proportion from 0.2 to 4 per cent, by dry weight. Therefore, the amount of ash left will depend on the tree species and which part of the tree has been burned – bark, trunk, branch, root or leaves. In addition, burn temperature can significantly affect the quantity and composition of the pile of residue, which is frequently less than the figures above due to flying ash.

The result is that most wood ash contains a high percentage of potassium and is recycled by gardeners who call it 'potash'.

Peter Gosling
Farnham, Surrey, UK

Swing low

While sitting on a playground swing, I have no reaction mass and nothing to push against (excluding the ground for the sake of this argument). I am puzzled because, given a small initial movement, I can increase the amplitude of the swing's movement by simply changing my posture on the seat at the peak of each swing. How?

Matthew Wiley
Huddersfield, West Yorkshire, UK

Energy change is the key to understanding this question. Swings convert potential energy into kinetic energy and back to potential energy with each swing. The greater the total energy, the larger the amplitude. So increasing the swing's amplitude can be achieved by either adding kinetic energy (for example, getting someone to push) or by adding potential energy (by raising and lowering your centre of mass at different points on the oscillation of the swing).

Changing your posture on the swing changes your centre of mass. If you raise your legs and sit upright towards the bottom of the swing's oscillation then the work done by your muscles to raise your centre of mass will add potential energy. So continual raising and lowering of your centre of mass with every oscillation will increase the amplitude of the swing.

This can be demonstrated easily by suspending a weight from a piece of string attached to a support that allows you to raise or lower the weight by pulling on the string. Set the weight swinging with a small amplitude. If you raise the weight slightly at the halfway point of the downward swing and lower it at the halfway point of the upward swing then, with a little practice, the amplitude of the swing will increase.

So what are you pushing against? When you change your posture to raise your centre of mass, your body pushes against the seat of the swing. This load is carried by the rope holding the seat to the swing's frame and so to the ground. So, even though you are not directly in contact with the ground, you are still pushing against it.

Gordon Gibb
Glasgow, UK

It is quite correct that moving the body in a circumferential direction will not excite a playground swing, because there is no reaction force against which the movement can do work and so transfer energy to the swing's oscillations. The way we were all taught as children to 'pump' a swing, by swinging our legs forward while moving forward and back while moving backwards, is very misleading because it appears to take place once per cycle, at the fundamental frequency. This back-and-forth motion transfers no energy.

The only way that energy can be fed into the swing from a body moving generally with it is by doing work in a radial direction against the so-called 'centrifugal' force. This force increases as the swing moves faster (at the lowest point of the swing) and is zero when it is stationary (at the highest point of the swing). In the traditional method of swing pumping, the energy is delivered by the small second-order radial movement of the centre of gravity, caused by the legs being hinged at the knees. This radial movement takes place twice per cycle, at the second harmonic.

From this, it can be seen that the traditional method of swinging is relatively inefficient. It is much more effective to stand on the swing, to bend the knees at each high (stationary) point, and to straighten them at each low (maximum velocity) point. This produces a large radial movement of the centre of gravity, again at the second harmonic. Once you get the knack of it, you can pump the swing up to terrifying heights in just a few cycles. At large amplitudes, it's easy to feel the extra work your knees are doing against the centrifugal force. Since

energy cannot be transferred to a stationary swing by this method, it is still necessary to have some initial amplitude, however small.

An interesting corollary, demonstrated to me by a friend, is that it is also possible to bring the swing nearly to rest within a few cycles – extracting energy from it by knee-bending of reversed phase, so that you are standing up at the high points and dropping down at the low ones.

The pumping of a swing is just one example of a wide class of what are called parametric phenomena, where some parameter (for example, moment of inertia, inductance, dielectric constant) of an oscillator is varied periodically at some frequency which is usually higher than its natural frequency. If the effect can maintain oscillations in the absence of any other input, the oscillations are called subharmonic. If the effect is weaker, but able to increase a continuous input signal, it's called parametric amplification. Both effects are used in modern electronic and optical systems.

J. B. Gunn
New York, US

On the double

If you travelled at the speed of light, would you get a light flash as with a sonic boom at the speed of sound?

Simon Williams
No address supplied

A sonic boom does not occur at the speed of sound; rather, it happens when the shock wave caused by an object going faster than sound passes the ear. Particles cannot exceed the speed of light in a vacuum, c. But if light is slowed to less than c, as it is in an optically dense transparent material, then it is easy to exceed the speed of light. When a charged particle exceeds the speed of light it does emit light in a way analogous to a sonic boom, which may be seen as a flash. This radiation was discovered by Pavel Cherenkov in 1934. Cherenkov radiation causes the blue glow in a water-moderated nuclear reactor.

Howard L. Medhurst
Crawley, West Sussex, UK

There is an analogue to the sonic boom for electromagnetic waves, but the particle must be travelling faster than the speed of light. This is possible because light travelling through a medium has a velocity less than its velocity in a vacuum. This lower velocity is given by $v = c/n$ where c is the speed of light in a vacuum, and n is the refractive index of the medium.

The radiation of light analogous to the sonic boom, called Cherenkov radiation, is produced whenever the velocity of a particle exceeds c/n. The blue glow that emanates from water in which highly radioactive nuclear reactor fuel rods are stored is caused by the Cherenkov effect. Much of the radiation that fuel rods emit is in the form of high-energy electrons. The electrons travel through the water at a velocity greater than that of light in water and hence cause the characteristic 'Cherenkov glow'.

The importance of the Cherenkov effect as a scientific tool lies in the connection between a particle's momentum and the angle at which the Cherenkov photons are emitted. A measurement of the angle of Cherenkov emission provides an indirect measurement of the speed and direction of a particle. Cherenkov detectors are one of the important tools used by particle physicists to probe the ultimate small-scale structure of matter.

Ray Hall
Illinois, US

Cherenkov radiation occurs when a charged particle travels through a refractive medium at a speed faster than the local speed of light. A refractive medium is necessary because nothing can travel faster than light in a vacuum. For particles of sufficiently high energy, even air has a high enough refractive index to produce Cherenkov radiation.

Pete Bleackley
Leicester, UK

Battered not scattered

The speed of light in air is about 0.997 of the speed of light in a vacuum. This means that light probably interacts with air molecules many millions of times, even over a short distance, and photons are scattered following collisions. Why then can we see images crisply and sharply? What kind of interaction slows light down and yet does not change its path?

Leo Sarasua
Rijswijk, The Netherlands

The slowing of light is an illusion, at least on a microscopic scale. Its speed between atoms is the same as its speed in a vacuum.

Light waves carry an alternating electric field that makes electrons move up and down in the same way that waves in water make boats bob up and down. In order to oscillate, each electron borrows some energy from the passing light beam. However, by virtue of their ever-changing speed and direction, these electrons are accelerating. This means they emit electromagnetic radiation. And, in fact, they emit exactly the same light they absorbed a moment earlier; it is just slightly delayed. This is why light appears to be slowed when not travelling in a vacuum.

Light can also be visualised as tiny particles of energy called photons. If a photon carries exactly the correct transition energy to promote an electron to a higher energy level, it will be absorbed and the medium will be perceived as opaque. For example, ozone absorbs ultraviolet light and so is opaque to it. This is why there is so much concern about ozone depletion and the increased skin cancer risk. Photons of visible light pass through the ozone layer, which is why we can't see it.

However, the closer a photon's energy is to an electron's transition energy, the longer it is delayed by each electron. This explains why the apparent slowing of light is frequency dependent, why we get dispersion of light, and also why blue light is slowed down more than red in glass.

Mike Follows
Willenhall, West Midlands, UK

When light travels in matter, it encounters atoms and molecules which are, in effect, immersed in a vacuum. Each atom interacts with the light and its electrons quiver in the light's electric field. As a result, each atom becomes a source of spherical waves emitted synchronously with the exciting radiation. Between any two interaction events light travels with its vacuum speed, but due to the myriad scattering its effective velocity is slower. In the visible part of the spectrum, light is slower in denser media.

In order to understand why the direction of propagation is not altered, the wave nature of light has to be taken into account. The spherical waves emitted from each atom interfere with the light emitted by all other atoms. The positions of the emitting atoms are generally random, but are all excited at different times, corresponding to their positions in space. It can be shown mathematically that in homogeneous media the only direction of constructive interference is the original direction of propagation. In other words, the

wavefront of the incoming light is duplicated and images are not distorted.

However, there are situations in which images are distorted in air because of convection, turbulence or temperature gradients. These phenomena are the main reason for placing and maintaining the Hubble telescope in the vacuum of space.

Iavor Veltchev
Huntingdon Valley, Pennsylvania, US

Push and pull

Why, in an atom, does the negatively charged electron not collapse into the positively charged nucleus? Is this in any way similar to the reason why large systems like stars and planets do not collapse into each other under the pull of gravity?

Daniel Doe
Northamptonshire, UK

When Ernest Rutherford, the New Zealand-born founder of nuclear physics, first discovered the atomic nucleus he did indeed propose that electrons did not fall toward the nucleus of the atom because the attractive forces of the nucleus were being balanced by the orbital velocity of the electron in much the same way as a planet orbiting a star.

However, the Danish physicist Niels Bohr modified this theory after Albert Einstein and Max Planck found that energy could only exist in certain discrete amounts, or quanta. This meant that electrons could be seen to have both wave and particle properties, and required that the circumference of the orbit of an electron could not be zero. This means, of course, it could never reach the nucleus.

We have since adopted the model proposed by the Austrian theoretical physicist Erwin Schrödinger. Instead of orbiting the nucleus like planets, his model has electrons occupying 'clouds' where it is statistically probable that they will exist, although we may never determine an electron's position and velocity at the same time.

Michael Ludlam
Sheffield, South Yorkshire, UK

Niels Bohr asked this very question in 1913. The atom was known to have a small heavy nucleus, and the much lighter electrons were thought to orbit it like planets around the sun. As long as a planet does not lose energy, it can continue its orbit indefinitely.

According to the laws of electromagnetism, charged particles moving in a circle ought to radiate energy as waves. Bohr calculated that a hydrogen atom should collapse with a flash of light in a matter of femtoseconds. Because this does not happen, he proposed what has become known as the 'old' quantum mechanics. It asserted that the electron's angular momentum had to be a multiple of Planck's constant.

The rule meant that electrons could only occupy particular orbits, and there was a minimum size of orbit. Using this, Bohr was able to predict the entire spectrum of excited states of hydrogen, which was a quite astounding achievement.

But Bohr's theory was hard to apply to more complex atoms and was superseded by Erwin Schrödinger's wave mechanics in 1927, which is the start of modern quantum theory.

Schrödinger's formulation shows that an electron has a wave character, and a stable atom can be thought of as a box confining the wave. An electron has a wavelength equal to Planck's constant divided by its momentum, so the faster an electron moves, the shorter its wavelength. To confine the electron near the nucleus the electron must move very quickly.

Conversely, a fast-moving electron can escape the pull of the nucleus. So you can think of the size of an atom as resulting from a compromise between the electrons having enough kinetic energy for their waves to fit in the box, but not so much that they can escape.

David Barnett
Institute for Advanced Physics
London, UK

As Niels Bohr realised in 1913, the electron just doesn't.

And no, large solar systems don't not collapse for quantum-mechanical reasons. They don't collapse because the planets' velocities keep them in freefall.

Jay M. Pasachoff
Field Memorial Professor of Astronomy
Williams College
Williamstown, Massachusetts, US

Hot metal

My son has a small toy car made of a die-cast metal. When it is immersed in icy water, the body paint turns purple. When put into warm water, it turns blue. If half-dipped it will show both colours. He has also worked out that if he breathes heavily on it, it will warm up and turn blue this way. How does it work?

Peter Bryant
London, UK

This is thermochromism – the ability of a substance to change colour when its temperature changes. A mood ring, which alters colour as the wearer's body temperature changes, is an excellent example of this, but it is more usefully incorporated into items such as baby bottles, which change colour when the milk inside is cool enough to drink.

Thermochromism can be based on liquid crystals but, in this case, a leuco dye has probably been added to the paint. This pigment switches between two states, depending on its temperature. In one state, it is transparent; in the other, it absorbs light at particular wavelengths. The absorbed colour is subtracted from the light reflected, giving the object a different colour.

Mike Follows
Willenhall, West Midlands, UK

Sideways glance

The health and safety message on the wall above our office microwave includes the advice that we should not look closely into the microwave while it is heating our food. What would happen if we ignored this advice? And how close is it safe for me to get?

Tracey Syrett
London, UK

We can see food cooking inside a microwave oven through the metal mesh incorporated into the glass of the door. This is because visible light has a wavelength of around 500 nanometres – about 5000 times smaller than the holes in the mesh – and so passes through unhindered.

But as far as microwaves are concerned, the mesh might as well be an impenetrable sheet of metal. This is because the microwave radiation used in ovens has a wavelength of around 12 centimetres, about 60 times the size of the holes in the mesh. It is perhaps easier to visualise the situation if we replace the waves with their particle equivalents, photons.

Photons of light are skinny enough to squeeze through the holes in the mesh whereas fat microwave photons are not.

However, when any electromagnetic radiation is reflected in this way, an 'evanescent wave' extends slightly beyond the reflector. The intensity of this radiation falls off exponentially and is virtually zero by the time it reaches the outer surface of the glass.

If you hold your hand against the window, you will not expose your skin to significant microwave radiation. But if a bigger hole were torn in the metal mesh, the evanescent wave would extend further out and you would risk burning your hand. A bigger risk comes from microwaves leaking through any gaps around an ill-fitting door.

Unless there were an obvious manufacturing fault or subsequent damage to the oven, it would probably be safe enough for you to stick your nose up against the glass. But why take an unnecessary risk? I suspect the message is a potential defence against a lawsuit in what is becoming an increasingly litigious society.

In defence of your employer, it might be unreasonable for them to keep a constant vigil on the microwave oven to ensure it is fault-free, but it might be helpful if the message urged users to report any damage, particularly to the metal mesh or the door seal.

Mike Follows
Willenhall, West Midlands, UK

4 Meteorology

Gust quest

What mechanisms are responsible for causing the wind to blow in gusts?

Chris Long
Sussex, UK

Near the surface of Earth, friction slows the wind. Turbulence is almost always created by layers of air moving at different velocities and this enhances or reduces the surface wind. The enhancements are the gusts. Strong turbulence is also created by obstructions such as buildings, which is why city centres are notoriously gusty.

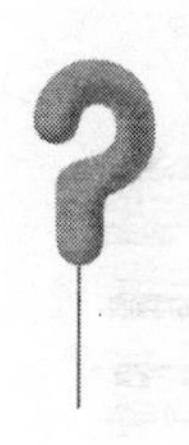

If the surface is sufficiently warmer than the air above, then convection will produce columns or walls of warm air called thermals. These will rise from the surface, and draw in currents of air to the base of the rising column. These currents can add to the mean wind to produce gusts that are longer lived than the usual turbulent gust.

In addition, if the convection is strong enough, it may produce shower clouds by condensation of moisture in the thermal as it rises and cools. Subsequent evaporation can then result in columns of cold air rapidly descending from these clouds to produce violent gusts at the surface. These are sometimes called squalls.

Mike Brettle
Cardington, Bedfordshire, UK

Cloud line

Why do most clouds have defined edges (or at least appear to)?

Richard Booth
Lewes, East Sussex, UK

The clouds with the most clearly defined edges are the billowing white cumulus clouds rising into a clear sky. These are formed by the condensation of water vapour as air expands and cools. This does not occur in a homogeneous layer but in a discrete parcel or column of warmer and less dense air rising from below through colder air above.

Although the cloud is cooled by expansion as it ascends, it continues to rise as long as its temperature is higher than that of the air surrounding it. Only when the air forming the cloud reaches a level at which the surrounding temperature is the same does it mix with that air and become fuzzy. Until then there is a sharp boundary between the different air masses.

Barrie Watson
Shoreham-by-Sea, West Sussex, UK

Clouds may appear to be static objects. In fact they are dynamic and there is usually a convection current of air rising up into a cloud. As this air rises, it expands and cools. At a given altitude, condensation occurs, thus defining a sharp lower boundary to the cloud. The sharpness of the upper surface depends on how fast the air is rising and the extent to which turbulence mixes this damp air with the surrounding, drier air. When the convection current ceases, the cloud will tend to become diffuse and lose its well-defined edge.

Glider pilots use the appearance of the clouds above them

in order to recognise where there are thermal currents that will enable their planes to gain altitude.

David Shirtliff
Tamworth, Staffordshire, UK

Wind up

Why is it more windy in winter, when the input of solar energy is least?

D. J. H. Wort
Whitby, North Yorkshire, UK

It is not the local input of solar energy that determines how windy it is, but the differences in heating of Earth's surface between cold polar regions and hot tropical regions. This leads to a circulation of air being set up to even out these differences, with general rising of hot air near the equator being replaced by air from near the poles.

If Earth did not spin, this would lead to a wind at Earth's surface blowing from the north in the northern hemisphere and vice versa in the south. However, due to Earth's rotation, the wind gradually curves to the west (the Coriolis effect) and air cannot move directly from pole to equator in one step. The single convection cell splits into three, with rising air not only at the equator but also in the temperate regions, and subsiding air in the subtropical desert regions, as well as in the polar regions. Thus, there is an area of conflicting winds in the temperate regions (between approximately 40 and 60° north and south).

In the northern hemisphere temperate zone, cold north-east winds battle with warm south-west winds. The strength of these winds is less in summer because Arctic land masses

are warmed by near continuous sunshine, and the difference in temperature across the temperate zone is not particularly marked. In winter, when the pole is in near continuous darkness, areas near it become very cold. Temperature differences across the temperate zone become large, strengthening vertical air currents in the convection cells, and making surface winds much stronger.

This is less noticeable in the southern hemisphere. Southern polar regions are land covered by ice (Antarctica) or ocean, neither of which is warmed much in summer. This causes the temperature differences across the temperate zone to be much the same all year.

Graham Hughes
Sutton, Surrey, UK

In winter, our weather is affected predominantly by depressions which are formed where warm, moist air, heated in the tropics, and cold, moist air from the North Pole meet over the Atlantic. The warm air rises above the cold. Both air masses spiral upwards and anticlockwise, forming a low-pressure area, the energy being supplied by latent heat stored in water vapour in the air masses.

Air from surrounding higher-pressure areas flows into the depression to reduce the pressure difference, causing wind.

Joe Walton
London, UK

Wind velocity in the UK is determined by the interaction of continental and North Atlantic air masses, rather than the solar heating of the islands. Strong winds are associated with the passage of small, deep depressions, which result from the mixing of warm Atlantic air (Gulf Stream) and cold polar air. They drift eastwards with the prevailing Atlantic airflow.

As Europe cools in winter, air pressure rises and a

macroscopic high forms over the continent. In the northern hemisphere, air circulates clockwise around a high.

This geostrophic wind causes Atlantic depressions to track northwards across the UK. As the depressions approach, their own circulation is augmented by the geostrophic wind, so strong westerly winds dominate.

In summer, the continent heats up, producing a low-pressure region over Europe.

The geostrophic airflow over the UK becomes northerly; approaching depressions track further south and are weakened by the countercurrent of the European cyclone and strong winds are less frequent.

Alan Calverd
Bishop's Stortford, Hertfordshire, UK

Dry seas

Over dinner recently, a friend mentioned what he called 'deserts at sea'. He was referring to areas of the world's oceans where rain seldom, if ever, falls. I can see how this might happen, as a lot of cloud formation and rainfall depends on topographical features such as mountains, which the sea obviously does not have. But do such deserts at sea really exist, how permanent are they, and have they been charted?

Sean Williams
Adelaide, South Australia

Arid areas of ocean certainly exist. They are caused by the circulation pattern of Earth's atmosphere.

In each hemisphere, between the latitudes of about 25 and 45°, there is a zone known as the subtropical high-pressure belt. This belt contains several separate high-pressure cells

(also known as anticyclones), and moves north and south depending on the season, being about 5° closer to the equator in winter than summer.

This zone is the subsiding arm of the Hadley cell, a north–south circulation of air in the low latitudes, consisting of two opposing cells, each having air rising in the intertropical convergence zone (around the equator) and sinking in the subtropical high-pressure belt.

In general, the belt comprises vast areas of light winds and gently subsiding air. The subsiding stable air undergoes compressional warming, producing low relative humidities. The weather is usually fine and rain clouds few, resulting in arid climates over both land and sea. However, the far-western portions of the subtropical highs have less subsidence and the air is not as stable, so cloudy, stormy weather is more frequent there.

Most of the world's great deserts lie under the eastern flank of the subtropical anticyclones: the Sahara, the Kalahari, the deserts of the Southwest US and the Atacama in Chile, as well as vast areas of inland and western Australia. Large areas of ocean in the subtropical high-pressure belt are also arid.

These zones became known as the horse latitudes during the days of the great sailing ships. Sailing ships becalmed in the belt would run short of water and, with no rain, horses being traded between Europe and the Americas would be thrown overboard, often into the Sargasso Sea.

Katrina McDonnell
Hornsby Heights, New South Wales, Australia

Large cone please ...

Why do tornadoes have the shape of an inverted cone?

Brian LaBelle
Brea, California, US

Exactly how tornadoes form is not completely understood, but we know that they usually occur during thunderstorms and when volumes of air are unstable. They are the result of updrafts that are created when warm, moist air meets cold, dry air. First, a horizontally spinning column of air called a vortex can form when there are different wind speeds at different altitudes. If this vortex collides with a violent updraft, it can be knocked into an upright position; when the vertically spinning vortex reaches the ground, a tornado is born.

To understand why tornadoes are often shaped like inverted cones, you need to remember that air pressure decreases at higher altitudes. Near ground level, the air surrounding the vortex is at high pressure and crushes the spinning column of air. As the altitude increases, the pressure outside the tornado drops, and it will spread out.

There is more than one way for a tornado to reveal itself, though not all are visible. At the centre of the tornado, the air is spiralling upwards at a great speed and creates a small region of lower pressure. If the pressure there is low enough, water vapour in the air will condense into visible droplets and the tornado will appear as a funnel-shaped cloud.

Sometimes, if the air pressure inside the tornado is not low enough for clouds to form, the tornado will only be revealed by the dirt and debris it picks up from the ground. On the other hand, if the air is very moist and the air pressure extremely low, then the cloud base may be so close to the

ground that there is not enough vertical distance for the funnel cloud to taper into a point and the tornado will take the form of a wedge.

Paul Knightley
Senior Meteorologist
PA WeatherCentre
London, UK

Anybody interested in seeing some tornadoes should check out www.chaseday.com/tornadoes – Ed.

Winds of change

Having witnessed the power and longevity of hurricanes in North America, I am wondering whether, in theory, a hurricane could continue for an indefinite length of time and could grow until it engulfed all other weather patterns and become the only weather system on Earth?

Simon Wallett
Nottingham, UK

No. Hurricanes depend upon a continual supply of warm water to maintain their energy and induce the low air-pressure environment in which they thrive. Their nemeses are cold water and high-pressure areas. In fact, hurricanes are only present between May and November, as it becomes too cold after this time.

Also, land masses, particularly mountains, disrupt the airflow to such an extent that the winds cannot coalesce into the fixed form of a hurricane system.

As powerful as they are, hurricanes are no match for this combination of factors. Their finely tuned structure cannot persist for long before something disrupts it. It is very rare for them to exist for longer than two weeks.

Bill Barnes
Warrington, Pennsylvania, US

Weather systems such as hurricanes require a source of energy – a warm ocean in this case. This warms the air above, so it becomes less dense and rises, leading to the reduction in pressure associated with the eye of these storms. Although they can be very destructive, they begin to dissipate once they make landfall because they are isolated from their energy source.

Hurricanes belong to a family of tropical cyclones that includes typhoons. The longest-lived of these was Typhoon John, sustained by the Pacific Ocean for 31 days during August and September 1994.

In contrast, the Great Red Spot on Jupiter is a storm that has been raging for more than three centuries. It was first observed by Robert Hooke in 1664 and its diameter is double that of Earth. Part of the secret of the storm's longevity is that Jupiter is a gas planet – there is no land over which it can stray.

Even if storms on Earth could reach continental dimensions, they would not be able to stray out of the hemisphere in which they develop. The equator essentially splits the globe into two independent air-circulation systems called Hadley cells. These are convection currents created by air rising at the equator and sinking at higher latitudes.

Mike Follows
Willenhall, West Midlands, UK

Hurricanes need a constant source of warm water to sustain their growth. The moisture from the ocean is fuel for a constant building of thunderstorms between the outer edges

and the eye of the hurricane. These thunderstorms release huge amounts of latent heat, which in turn gives more energy to the hurricane.

For a hurricane to continue to grow, its outflow needs to stay vertical. This is nearly always hindered by the normal east to west jet streams that are present in the atmosphere.

The answer to the question of whether a hurricane could grow 'until it engulfed all other weather patterns' would be no. However, if there are no strong upper air streams and the ocean remains warm, it is certainly possible for hurricanes to become more powerful.

The key is that the warmer the oceans are, the warmer the air will be and consequently the more capacity the air has for holding water vapour. This would allow stronger thunderstorms within the hurricane itself.

David Saunders
Lecturer in Meteorology
University of Maryland, US

Strike action

If I am swimming in an outdoor pool surrounded by tall trees, and a tropical thunderstorm breaks out, am I at an increased risk of being struck by lightning in the pool? And will it harm me?

Simon Hare
Siem Reap, Cambodia

Lightning normally strikes tall objects, particularly when they stand on an otherwise flat and featureless landscape.

The base of a storm cloud is generally negatively charged, which, by repelling electrons, induces a positive charge on the ground below. An electric field is thus set up in the air sandwiched between the cloud and the ground. Lightning will take the path of least resistance but, if the cloud base and the ground are both flat, there is no obvious route for an electrical discharge.

Given that a tall object should increase the local electrical field strength, one might naively expect surrounding trees to act as decoys for a lightning strike. However, there are many other variables, not least the shape and height of the cloud base.

Water is a good electrical conductor so, when lightning strikes, the current tends to be confined to the surface and spreads out in all directions. This puts swimmers at risk and there are plenty of documented incidences of injury and death. For example, two out of a group of nine swimmers were seriously injured in waters off Chiba Prefecture, Japan, in 2005, after a lightning strike. Perhaps they were far enough away from the point of impact to escape with their lives.

A typical swimming pool might not be sufficiently big to dissipate the energy associated with a lightning strike. In July 2006, Michael Haffenden was standing by the metal steps of the pool of a hilltop villa he had rented in Tuscany, Italy, for a family holiday. Tragically he was killed when lightning struck the pool, even though there would undoubtedly have been taller objects close by. Given that lightning is known to travel through plumbing, some experts even recommend staying out of indoor pools, baths or showers during electrical storms.

Leaving the pool is the best option when a storm is approaching. The trees would probably reduce the chances of the pool being struck but, given the finite size of a pool, a swimmer would be too close to the strike to avoid serious injury or death. If a storm arrived out of the blue and I could feel my hair standing on end, I would be tempted to duck

under the water and swim submerged to the edge of the pool, as far away as possible from any ladders that extend above the poolside.

Mike Follows
Willenhall, West Midlands, UK

Crunch time

Why does freshly fallen snow squeak and creak when you step on it?

Pauline Lacey
Lincoln, UK

The sound of feet on snow is an example of the stick-slip phenomenon. Other examples are squealing tyres and violin music.

When you try to drag one object over another the friction between them prevents movement. The objects won't move at all until the dragging force at least matches the friction force. However, if they are elastic (and everything is a bit elastic), they will stretch. Then, when the elastic force matches the friction force, sliding starts.

For most materials, the friction force is higher when they are not moving than when they are. When movement starts, the friction force drops and the stretched elastic force will suddenly be too big, so the dragged object will accelerate as the stretch contracts. The object's momentum will carry it past the point at which the elastic force is less than the friction force, until eventually it stops again.

Now the higher static friction comes back into play and the objects stick together until the elastic force once again matches the higher static friction force. Then the cycle repeats.

The result is a sawtooth vibration of gradual stretching and sudden release – and this provides the creaking noise of snow. In reality, there is not a step transition between the two types of friction; the friction force just reduces very rapidly as relative movement starts.

There are a number of simultaneous microscopic phenomena that give rise to this reduction, such as the behaviour of any thin films of liquid between the surfaces. And, as rough surfaces start to slide, the collisions between their minuscule hills and dales tend to throw them apart slightly, reducing friction between them.

In the case of snow, I suspect that the principal cause of the sudden reduction in friction is the famous pressure-melting of ice that allows skaters to get such low-friction sliding. A build-up of pressure as the snow is elastically compressed underfoot finally causes a tiny bit of it to melt, suddenly reducing friction and allowing the stored elastic energy to dissipate in movement. As the movement reduces the elastic pressure, the ice refreezes, and the cycle starts again.

Not all materials behave in this way. It is virtually impossible to get polytetrafluoroethylene or PTFE (the coating on non-stick frying pans) to stick-slip because, unusually, its dynamic coefficient of friction is a little higher than its static one. But the next time a squeaking knife on a plate sets your teeth on edge, you can blame the difference between static and dynamic friction.

Adrian Bowyer
Bath, Somerset, UK

It was a very snowy start to the year all across Canada, so there are probably 35 million Canadians who can now answer this question.

From personal experience of walking in fresh snow at −37 °C, I have found that you can tell the temperature from

the sound of the squeak. The squeak gets higher in pitch as the temperature falls.

It is all about the shape of the snowflake clumps. On the Canadian prairies where it is extremely cold in January and February, the snow is often so light and uncompressed that you can blow it off your clothes after it has fallen on you. The density is so low that lifting a large shovelful to clear your driveway is not especially arduous.

Around the Great Lakes and Toronto, the air temperature remains higher and the snow clumps together much more easily. There, a shovelful is much heavier – heart attack-inducing unless you use a smaller shovel. In April and May there is also very wet snow on the prairies, so here you must make sure that you only move a quarter of a shovel of snow at a time. All of this, I'm pretty certain, also explains why western Canadian snow shovels are wider than those from eastern Canada.

Another explanation for the squeak is that at low temperatures you crush the snow when you step on it, causing a squeak, rather than pressure-melt it. However, this doesn't explain everything because the squeak disappears when the snow has remained at –37 °C for a long time. I have always considered this to be associated with the snowflakes subliming – evaporating without becoming a liquid – and becoming more rounded rather than pointy. If rounded, the flakes merge together very well without having to melt.

Mike Smith
University of Calgary
Alberta, Canada

Rain, rain, go away

The chair of the UK's Environment Agency, Lord Smith of Finsbury, has been warning of 'a new kind of rain'. Apparently, 'convective rain' does not sweep across the country but dumps its deluge in just one place, putting great strain on that particular area. Is this a real and new phenomenon caused by climate change? And if it is, what causes it?

Pavel Myres
Birmingham, UK

Convective precipitation is by no means a new phenomenon. Indeed, every time the phrase 'sunshine and showers' is used in weather forecasts, convective rainfall is implied. It is caused by a sharp temperature gradient in the lower atmosphere.

Warm air at the surface of Earth rises through cold air higher in the atmosphere in columns known as thermals. As a thermal rises it cools, and the moisture contained within it condenses to form towering cumulonimbus clouds. These can then deposit heavy but short-lived bursts of rain or snow.

In summer, the sun can heat the air just above the ground so much that huge thermals rise to give us thunderstorms.

In winter, during a cold snap, warmer air over the sea can gather lots of water and rise into a bitterly cold air mass, causing intense falls of snow. This is sometimes called 'lake effect' snow, so named after a similar phenomenon over the Great Lakes of North America which gives severe blizzards in winter.

Convective precipitation is very different from frontal precipitation, which is caused by the sideways collision of two air masses of different temperatures forcing warm air to rise and resulting in thick cloud.

Slow-moving convective precipitation can be caused by 'convergence zones', where two warm air masses meet and are forced upwards. This was responsible for the severe flooding in Boscastle in Cornwall, UK, in August 2004.

Although this is not a new phenomenon, warmer air in summer caused by climate change perhaps makes these events more likely.

William Torgerson
York, UK

Paindrops keep falling

This morning it was raining so hard the drops were painful. What causes this? Is it drop size and the height from which they fall? Or is there another mechanism that increases the speed at which drops fall? The rain seemed to be coming down especially hard, and drops hitting the ground bounced high into the air.

Paul Freyer
Andover, Hampshire, UK

Two things affect the energy of a raindrop and hence how painful it feels when it hits your skin: its size, and the speed at which it falls. In practice, raindrops cannot grow bigger than about 5 millimetres in diameter, because air resistance will cause drops above this size to disintegrate into smaller ones.

Now, the larger the raindrop, the higher its maximum speed, or terminal velocity. When it reaches this speed, air resistance balances the downward force of gravity, so the drop no longer accelerates.

A 5mm-diameter drop rapidly reaches its terminal velocity of around 9 metres per second, depending on the prevailing air temperature and humidity, so the height from which a raindrop falls has no bearing on its pain-inducing ability.

The school in which I teach is housed in a tall building with narrow stairwells where there is minimal horizontal air movement – ideal for investigating the behaviour of 'raindrops' we make.

We find that a 4mm-diameter drop falling two floors into a puddle produces a splash or 'bounce' of about 50 millimetres. The bounce for one falling four floors is about 150 millimetres, and for one falling six floors, just over 200 millimetres. This last result is no surprise as drops probably reach terminal velocity after falling about five floors.

Our experiments show that a drop 4 millimetres in diameter falling six floors on to skin feels like being hit by grains of rice from a few metres away. Drops that are 5 millimetres in diameter have slightly greater terminal velocity, but they are also almost twice as heavy. This means they have about twice as much energy and it feels mildly painful when falling on to skin from six floors up.

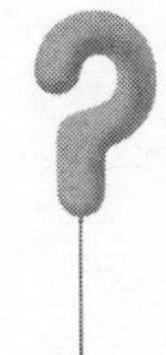

Thanks to bald colleagues for their assistance.

David Muir
Science Department
Portobello High School
Edinburgh, UK

Alarming noises

Why do heavy thunderstorms and fireworks set off car alarms?

Mike Preston
London, UK

The alarm systems of modern cars have a variety of sensors, but the ones that are most likely to react to thunderstorms or fireworks are shock and pressure detectors. A shock detector

senses if something bumps your car in some way. A pressure detector senses a change in air pressure inside the car if a door is opened or a window is knocked in.

The speakers for a car stereo can act as pressure detectors. A speaker normally works by vibrating, according to the input of electrical signals, to produce sound. When used as a detector, the opposite happens. A loud sound or change in air pressure moves the speaker, producing an electrical signal that can trigger the alarm.

Lightning, or an explosion from a firework, produces a shock wave that propagates radially outwards from its source. A shock wave is a compression wave, characterised by a sharp increase in air pressure followed by a sudden drop. If the initial detonation is energetic enough, the shock wave expands supersonically and is heard as a loud, truncated clap or click. When such a wave hits a solid body, some of its energy is absorbed, deforming that body, so it can easily set off a car alarm.

Shock waves can do more than set off car alarms – they can kill. Waves from high explosives can make a skull distort, causing concussion and brain damage. Many soldiers in the First World War who were close to shell impacts were killed, even though they showed little outward sign of damage. And the shock waves from roadside bombs can cause traumatic brain injury and subsequent mental health problems. This is a hidden legacy of the recent campaigns in Iraq and Afghanistan.

David Muir
Science Department
Portobello High School
Edinburgh, UK

A comprehensive modern car alarm system usually has a wide range of sensors. These include devices that monitor electrical activity within the car's own circuitry and acoustic and electromagnetic phenomena in the immediate vicinity, and perimeter radars and tilt-sensors. These sensors check for indicators that, individually or collectively, may suggest an attempt to steal, break into or damage the car. A central processing unit then cross-references the data to decide whether to activate the alarm.

The oldest types of sensors, essentially microphones (although some acoustic monitors now use the speakers on a car's sound system), listen for the distinctive frequency of glass cracking, which might indicate a direct or clumsy method of entry to the car. Electrical sensors detect anomalous voltage drops, such as those caused if a light comes on because the doors have been opened even though they've not been unlocked correctly.

Perimeter radar detects approaching objects, and the central processor determines whether they pose a threat according to set criteria. Then there are electromagnetic sensors that monitor sudden variations in air pressure, such as if a window had been shattered or a door forced. Last, the tilt-sensor, usually a float suspended in a mercury-filled capsule, monitors the inclination of the car, in the same way as the fluid in the inner ear regulates our sense of balance. This alerts the car to any attempts to tow it or load it onto a trailer.

So in a thunderstorm, it might be possible for fluctuations in atmospheric pressure to trip the sensors, and for lightning to fool the proximity radar or cause a power surge in the car's circuitry. It is not obvious how fireworks or a crack of thunder might do this, however, and the likely reason is that the noise fools the acoustic pick-ups into thinking that the shock wave they are detecting is the sound of breaking glass.

That said, it is by no means impossible for sophisticated car alarm sensors to be duped, albeit under bizarre circumstances. When the Eurofighter Typhoon first entered service with the UK's Royal Air Force and began attending air displays, commentators would warn the public that when the aircraft made a low pass its electronic countermeasures system could set off nearby car alarms.

Hadrian Jeffs
Norwich, Norfolk, UK

5 Chemistry

Free the atoms

Oxygen has a slightly greater density than nitrogen. Why don't these main constituents of air separate out?

Gerald Leach
London, UK

Gas molecules move rapidly at room temperature, with oxygen and nitrogen travelling at around 500 metres per second, so they obviously collide frequently. This allows the oxygen and nitrogen molecules to mingle and mix, rather like large numbers of people on a nightclub dance floor, in a process known as diffusion. Convection, the transfer of heat within the atmosphere, also plays an important role in gas mixing.

Gas mixing is a spontaneous process. This means that if you had a container with two compartments separated by a barrier, with one compartment containing pure nitrogen and the other pure oxygen, the two gases would automatically mix as soon as the barrier was removed.

Kenneth Koon
By email, no address supplied

A change in the ratio of oxygen to nitrogen would be expected in a hypothetical quiescent atmosphere. However, constant mixing occurs in the real atmosphere, driven by Earth's rotation and also by differences in density between hot air at Earth's surface and colder air higher up.

Up to altitudes of between 80 and 120 kilometres this mixing results in a fairly uniform concentration of oxygen and nitrogen – which respectively make up approximately 21 per cent and 78 per cent of the atmosphere.

This region is known as the homosphere. Partial stratification of the two gases does occur above 120 kilometres, in the heterosphere, where the density of air is much lower than at the surface and the efficiency of bulk mixing processes is reduced.

Simon Iveson and Mark Hentschel
Alumni of University of Newcastle
New South Wales, Australia

If there were no circulation in the atmosphere, the oxygen would tend to concentrate in the lower strata. This process would take millions of years once circulation ceased because molecules of oxygen (and, indeed, nitrogen) are constantly colliding with other molecules, meaning that it would take a long time for a particular molecule to fall from its starting point to the ground. Once it hit the ground, it would bounce and eventually rise again to a great height, only to fall again. This would be repeated frequently if no other variable, such as temperature, changed.

Although the individual molecules continue to travel up and down, each 'species' of oxygen and nitrogen would eventually reach an equilibrium distribution of molecules per unit volume as a function of height. This species density will decrease with height by an amount that depends on the molecular weight of the species. So the oxygen would fall off with height slightly faster than the nitrogen. At the surface, this would give a slightly higher concentration of oxygen than we experience now. At high altitudes, the air would become richer in nitrogen, but then other gases such as water vapour, neon, methane, helium and hydrogen would dominate.

In fact, atmospheric circulation and turbulence prevents this from happening in the lower atmosphere. But in the very high atmosphere there is not much circulation and the composition does indeed become dominated by atomic oxygen. Above 600 kilometres this is superseded by helium, and eventually by atomic hydrogen.

Eric Kvaalen
Paris, France

The average speed of an oxygen molecule at 27 °C is calculated at 484 metres per second. If you mix in slightly lighter nitrogen molecules those will move even faster. Even at those speeds, the molecules will collide with each other after travelling only about 68 nanometres. This gives the molecules no chance to settle.

It is like entering a classroom full of teenage boys and girls on the last day before the summer break and patiently expecting them to settle down to serious work. It won't happen.

Sjoerd Spoelstra
Physics teacher
Rotterdam, The Netherlands

Hidden gas

Where do neon light manufacturers get their neon from? I know from school that any neon present on Earth was originally made in supernovae. But where is it now? Being inert, it cannot form compounds from which we can extract it, so is it just floating about in the atmosphere and, if so, how do we isolate and concentrate it for use in light tubes?

Fabian Nesmith
Pahiatua, New Zealand

Neon certainly is the stuff of stars, but not just of supernovae: most bright stars make neon and eject some of it from their coronae.

On Earth, practically all our neon comes from traces once trapped inside the solid mass of the planet and now escaping continuously through volcanism and sea-floor spreading. Neon atoms are too massive to diffuse rapidly out into space as helium does, so they loiter permanently in Earth's atmosphere, as do argon, krypton and xenon.

Neon amounts to less than 20 parts per million of the atmosphere – equivalent to about a cubic metre in a 30-storey office block. But even at that concentration, it is the sixth most abundant gas present in the atmosphere, and our most abundant noble gas apart from argon. Fortunately, it is not very difficult to extract: the liquid-air industry produces tonnes of neon in the same way it does the other noble gases – by fractional distillation. Neon distils out first because it is the most volatile of the atmospheric gases apart from hydrogen and helium, which do not liquefy under the usual processing conditions.

Once extracted, neon is concentrated by the same means that we are all best made to concentrate: by being put under pressure.

Jon Richfield
Somerset West, South Africa

Neon gas occurs in tiny quantities in Earth's atmosphere and is trapped in equally small quantities within the rocks of Earth's crust. Dry air contains only 0.0018 per cent neon by volume.

The gas is produced industrially by fractional distillation of liquid air. Neon separates in the most volatile fraction, which also contains helium and nitrogen. The nitrogen is removed by condensation under high pressure and reduced temperatures and is absorbed into highly cooled charcoal. The helium is separated by selective absorption on activated charcoal at low temperatures.

However, because the atmosphere contains only minute amounts of neon, processing 40,000 kilograms of liquid air will produce only around 0.5 kilograms of neon.

Not all the lights we call neon lights actually contain the gas. Some contain mercury vapour at very low pressure and are coated on the inside with materials that fluoresce when they absorb ultraviolet radiation produced by passing an electric current through the mercury vapour.

Zoë Monnier-Hovell
Cambridge, UK

Plant poser

Did all the oxygen in Earth's atmosphere come from photosynthesising plants? If not, where did it come from?

Gabby Griffith
Cardiff, UK

In Earth's crust, oxygen combines with all the most common atoms to form water, rock, organic compounds and almost everything around us. Spontaneous free oxygen is about as likely as finding round rocks perched on steep slopes. Such rocks would imply that something had pushed them uphill more strongly than they could roll downhill.

Similarly, any free oxygen about us has been torn from its compounds with more than its bonding force. And that is a lot of force that only a few things are able to produce. Ionising radiation, such as X-rays, can do it, but there is little of that about. Visible light does it laboriously, step-by-step

through photosynthesis, the only process that could release the breathtaking amount (no pun intended) of oxygen that we see about us. On the back of an envelope I've calculated this to be perhaps 10^{15} tonnes.

How much oxygen plants actually produce is another matter. The chloroplasts used by plants to photosynthesise are thought to have originated as symbiotic cyanobacteria. So, in effect, all our oxygen came from photosynthesising bacteria.

Antony David
London, UK

Practically all of the atmospheric oxygen is of biological origin. The main culprit, however, is not plants but humble cyanobacteria. These single-cell organisms, which were present on Earth more than 3.5 billion years ago and pre-date plants, were initially responsible for all oxygen production and are still responsible for more than 60 per cent of current oxygen production.

Cyanobacteria come in many varieties and are sometimes called blue-green algae, although they are not really algae. A species of cyanobacteria present in the ocean, *Prochlorococcus marinus*, is both the smallest photosynthetic organism known and the most abundant of any photosynthetic species on the planet. It was only discovered in 1988.

Elmars Krausz
Canberra, ACT, Australia

The heat is on

What is fire made of? What is its atomic structure, what causes things to burst into flame in the first place and why can't all materials be made to produce flame?

Paul Heard
Rugby, Warwickshire, UK

Fire involves a chemical reaction between fuel and atmospheric oxygen. Once initiated it is self-sustaining, generates high temperatures and releases a combination of heat, light, noxious gases and particulate matter.

The visible flame is the region in which this chemical process occurs and so flame is essentially a gas phase phenomenon. For flaming combustion to occur, solid and liquid fuels must be converted into gaseous form.

For liquid fuels this is achieved by evaporative boiling. For solid fuels, the solid is chemically decomposed through the process of pyrolysis to generate volatile gases.

Ed Galea
University of Greenwich, London, UK

A flame is a region containing very hot atoms. At high enough temperatures all atoms will emit energy in the form of light as their electrons, which have been prompted to higher energy levels by absorbing heat energy, fall to lower energy states.

Because this light is emitted in discrete amounts according to the relationship $E = hv$ (where E = energy, h = Planck's constant and v = frequency), flame colour is related to the magnitude of the quantity of energy which is transformed to light.

This can most easily be seen with a Bunsen burner. A Bunsen burner that has a choked air supply burns cool, the light emissions from carbon atoms are relatively low in

energy and appear more red or orange. However, when the Bunsen is allowed air so that combustion is complete, the flame is hotter and the light emitted is of a higher energy and frequency and appears blue.

The luminescence of a flame is only half the story. The structure of the flame region is important to understand too. The flame area in a normal combustion environment, such as an open-air bonfire, is structured by convection currents which form as hotter, lighter air rises and allows cooler fresh air to replace it. It is this channelling effect and movement of air that shapes the dancing flames. It is interesting that in space, in zero gravity, the hotter and cooler air cannot move by convection, so flames take on weird shapes and may be stifled by their own combustion products.

Roger Doonan
London, UK

6 Evolution

Five live

Why do we have five fingers on each hand and five toes on each foot?

Joe Baris Spring (aged 6)
Ankara, Turkey

Humans' remote tassel-finned or fringe-finned (crossopterygian) ancestors emerged from the water with a limb architecture of one bone from the shoulder or hip, two bones from the elbow or knee, and several bones from the wrist or ankle. All land vertebrates have limbs that are based on that original scheme, including humans.

These pioneers had lots of slender 'toes' on all four feet – too many and too slender for control and power. There must have been strong selection for a more definite joint structure and more strength in each digit. By the time the first true amphibians appeared, toes had thickened and been reduced to eight or so on each foot.

Long before the first reptiles evolved, five toes had become pretty much standard issue. Mammals continued the pattern, which seems to be so robust and versatile that it has persisted among most non-specialised groups and a good few specialists as well, such as tree climbers and their descendants, including humans.

Specialisation tends to reduce the number of toes. Creatures that run, for example, need light feet more than they need versatile bone architecture, so their toes reduce in number – down to one in horses – and in size – two main toes and a couple of vestigial ones in artiodactyls (cattle, deer and suchlike). Some creatures, such as snakes, have even lost entire limbs. The only example of an added toe that I can think of is the giant panda's 'thumb', which is, of course, not really a toe.

Jon Richfield
Somerset West, South Africa

Group therapy

Why have humans developed different blood types and is there an evolutionary advantage? Surely, if blood types are due to random mutations, one will be a better performer in general terms than the others? And why can't different types be used for all patients requiring transfusions?

Iain Cunningham
Fareham, Hampshire, UK

There are four blood types: A, B, AB and O. These designations refer to the types of sugars (A, B and O) found on the surface of red blood cells. Everyone on the planet has an O sugar, and those who have no other type are known as blood group O. The other group names arise from the fact that some people have A, B or both A and B sugars attached to the O sugar.

The genetics is unusual in that there are two equally dominant alleles. If the gene from your mother is the recessive allele for producing O cell surface markers, and the gene from your father is for A cell surface markers, then your overall blood type is A. Why? Since everyone has O we only look at the second sugar present to determine blood type. If the gene from your mother was for B and the gene from your father was for A, then your blood type would be AB. If both parents donated A alleles, then you would be A. And if both gave you recessive alleles then your blood type would be O.

Cells use things such as proteins and sugars on their surface for many purposes. One is to enable the immune system to tell 'self' from 'non-self' and distinguish 'you' from every 'foreign' body that may invade. Our cells have lots of different types of surface markers that tell the immune cells not only that they are self but also what type of cells they are. Red blood cells have the A, B and O markers and can also be rhesus positive or negative, depending on whether a separate marker is present or absent.

Why can't any type of blood be given to anyone else? Our immune system attacks anything that isn't recognised as self. That means that if I have type A blood, meaning that I have both A and O sugars on my red blood cells, I can accept both type A and O blood from a donor. My body recognises both A and O as self. If type B blood was given by mistake, my immune system would attack those blood cells, and the transfusion would kill me.

Type O blood is the universal donor, because nobody makes antibodies to this blood type – we all have the O sugar. Type AB blood is known as the universal acceptor because all blood sugars are recognised as self.

Mark Sullivan
Seattle, Washington, US

The ABO blood system provides us with an example of stable polymorphism (or 'many forms') because the alleles responsible for the four different blood groups occur in fairly constant proportions. Alleles A and O are more common than B, so that about 40 per cent of people are in group A, 40 per cent in group O, fewer than 20 per cent in group B and only 2 per cent in group AB.

These are global figures and there are some regional differences. For example, Western Europe is dominated by groups A and O but Celts, such as the Irish, are nearly 80 per cent group B. The Indian subcontinent also has a preponderance of B alleles.

At first sight these statistics are difficult to explain. Surely one blood group is as good as another? In fact, there is now strong evidence to indicate that blood groups confer protection or susceptibility to a wide range of human diseases. Groups A and AB are more susceptible to smallpox (thankfully eradicated), group A is associated with stomach cancer, while group O has an increased likelihood of developing duodenal ulcers.

The problem of blood groups and transfusions is related but different – it involves antibodies to a particular blood group. If a mistake is made in matching blood types, the patient's antibodies will treat the transfused blood as an invading infection and try to destroy the red blood cells it has received. Of course, this rarely happens unless a major mistake has occurred in medical operating procedures.

A more common problem happens in pregnancy with the rhesus aspect of the blood system, when a rhesus negative mother carries a rhesus positive child.

During the last month of pregnancy, fragments of fetal red blood cells containing the rhesus antigen cross the placental membrane into the mother's bloodstream; the mother responds by producing rhesus antibodies which later pass back to the fetus, destroying its red blood cells. This rarely does enough damage to affect a first child, but it sensitises the mother so that, if she conceives another rhesus positive child, her body will start producing antibodies much earlier in the pregnancy.

This condition, known as erythroblastosis fetalis, is likely to kill the child unless it receives a blood transfusion of rhesus negative blood while it is still in the uterus.

Prevention is now possible with an anti-rhesus globulin that coats the fetal cells and prevents the rhesus factor antigen entering the mother's blood in the first place.

Mark Abbott
Moncalieri, Italy

Certain blood types tend towards susceptibility to particular pathogens, but it does not follow that any type is superior to all others. Microbes vary and populations with just one blood type could be disastrously vulnerable to particular epidemics.

Additionally, types resistant to diseases from one region may be at risk elsewhere. The distribution of blood groups reflects this. For example, antigens known as Duffy antigens are relatively rare in African populations; they seem to be associated with susceptibility to *Plasmodium vivax* malaria. Also A-type blood seems to go with susceptibility to the debilitating waterborne disease schistosomiasis (bilharzia), and this is consistent with the distribution of type A blood in Africa. Such distributions are a more common selective outcome in evolution than a quick takeover by just one allele.

Antony David
London, UK

Smart evolution

Over the past few hundred thousand years, humans have greatly increased in intelligence and in the size of their brains. As intelligence appears to produce an advantage to survival, do all animals show a tendency to evolve greater intelligence over time? Or is the value of intelligence overrated?

Derek Wroe
Stafford, UK

The conventional answer is actually rather dull: animals are as intelligent as they need to be and no more. This is because brains are expensive to run – your brain uses up 19 per cent of your body's total energy consumption – so any animal with a large brain has to justify the cost. For example, over the past few hundred thousand years, cheetahs and antelopes have been locked in a constant arms race but neither has chosen to evolve a bigger brain or, presumably, greater intelligence. Even though it is one option open to their evolution, and hence their ability to capture or evade the other, evolution has selected not to exercise it.

Yet there does appear to have been an increase in intelligence over the past few hundred million years. Mammals are more intelligent than fish, fish are more intelligent than crustaceans. Why is this?

Bigger animals have bigger brains with more neurons. More neurons are needed to control their larger muscle mass and to collect data from their extra sensors. It seems that one side effect of having a bigger brain is to be more intelligent.

So it seems that evolution has led to increasing intelligence simply because it has led to increasing body size. Vertebrates can be bigger than insects and nematodes because, among other adaptations, they have internal skeletons and oxygen-carrying blood. Greater intelligence is a by-product. Yet, in spite of these considerations, it is clear that humans have a bigger brain than would be expected for a mammal of our size.

The traditional explanation is that our greatly increased intelligence gave us a survival advantage, but there's little evidence for that. If extra intelligence helped our ancestors survive, then we would expect evidence in the form of increasingly sophisticated tools. That doesn't seem to have happened. *Homo habilis*'s tools stayed exactly the same for a million years, then *Homo erectus*'s tools stayed the same for another million years.

One theory says that unusually high intelligence aids our survival as much as a big tail aids a peacock's survival. It costs a lot to maintain and so reduced our ancestors' life expectancy. But like a peacock's tail it made us sexier, and the sexiest is, of course, the most likely to reproduce. We chat each other up. We write sonnets. We tell jokes (scan any personal column to see the importance of GSOH).

Sexual selection is arbitrary: the object of desire can be bright scales, large horns or a big brain. And the positive feedback of sexual selection leads to fast evolution and great exaggeration of the chosen characteristic.

Peter Balch
Edinburgh, UK

Although it is true that you need a large body to support a large brain, the two don't necessarily go together – witness the dinosaurs, many of which had small brains and huge bodies. But, of course, they were very successful for much longer than humans are likely to be – Ed.

Although many animals have evolved quite a high level of intelligence, it is by no means a desirable feature for all species to possess.

Animals have a limited supply of energy, so they have to choose how it is allocated. Two major uses of energy are brain power and reproduction. Humans and many other intelligent species have only a few offspring and look after them very carefully, which is partly thanks to intelligence. In contrast, insects, for example, tend to produce millions of eggs in the hope that a few will survive. This is very costly in terms of energy and any evolutionary trend towards having a larger brain would mean that they would have to reduce the number of eggs they produce.

So there are alternative ways of living – being intelligent

with a low reproductive rate, having a simple brain and a high reproductive rate, or somewhere in between. It is simplistic, therefore, and a little arrogant, for humans to assume that intelligence is the be-all and end-all of evolution.

Andrew Pine
Cambridge, UK

Who needs men?

Why is the ratio of men to women roughly equal? As a man can impregnate a woman more quickly than a woman can make a baby, the human race could easily survive on a ratio of, say, 50 women to one man, so is there another reason for the equality?

Julian Harrow
Southampton, Hampshire, UK

Ronald Fisher, who pioneered the mathematical theory of natural selection, wrote in *Natural Selection, Heredity and Eugenics: Including Selected Correspondence of R. A. Fisher with Leonard Darwin and others* (edited by J. H. Bennett, OUP, 1983):

> There are a number of instances of tendencies which have developed apparently clean contrary to the general interest of the species ... I think a good example of this is in the sex ratio of polygamous animals living in flocks and herds, where the economy of the herd as a whole would seem to suggest (and the stock breeder would prefer) a sex ratio of about 5 per cent males, but where nature, through the action of selection, insists on producing nearly equal numbers of the two sexes.

Elsewhere he argued that in a population of 5 per cent males any male has 20 times the influence of any female on the genetic make-up of the next generation. This gives an enormous selective advantage to a gene that predisposes an animal to have male offspring. The spread of this gene will increase the proportion of males, but its advantage will decline as the proportion increases towards 50 per cent. This broad argument is still accepted by biologists.

The main development in thinking since has been to observe that selection should favour equal parental investment in offspring of the two sexes, not necessarily numerical equality. This means that if the cost of raising a son were twice that of a daughter we would expect a sex ratio of two (females) to one (male).

But other complications arise. For example, R. L. Trivers and D. E. Willard pointed out in 1973 that in many species the variation of male reproductive success is larger than that of females – most females are likely to reproduce but only a few males do so, namely the larger dominant ones (*Science*, vol. 179, p. 90). This means that a mother, in a healthy condition, is likely to have more grandchildren if she puts her abundant resources into raising large strong males. If she is less healthy then producing more daughters is a better bet. Even so, the overall investment in the sexes at population level is still expected to be about equal.

Murray Lark
Bedford, UK

Fisher gave the answer to this question early in the 20th century. If there is a minority of one sex (say males), then there is a selective advantage in being male.

Later analysis has shown that this equality actually applies to the effort put into creating and raising each sex. Therefore, if it takes more resources to raise a female than a male, then the numerical sex ratio will be biased to males. The age at which the sex ratio applies is when the offspring become independent of the parent.

An interesting paradox is seen in red deer. Tim Clutton-Brock, a professor of animal ecology, has shown that male calves are more of a drain on the mother. However, the sex ratio is biased towards males. This is explained by the fact that the males leave the maternal herd when weaned. However, the female calves remain and are therefore in competition with the mother, reducing her resources.

John Rostron
School of Health and Biosciences,
University of East London, UK

Food chains

Broadly speaking, plants obtain energy through photosynthesis of sunlight, and animals by eating those plants or each other. Surely evolution will have experimented with other means. Is there any fossil evidence to suggest what this might have been?

Brian Wall
Ferndown, Dorset, UK

Evolution may have experimented with primary sources of energy other than sunlight, but at present no authenticated trace of such organisms exists on Earth. Anaerobic photosynthesis seems to have underpinned life for effectively all of the early aeons of its terrestrial existence, before oxygenic photosynthesis emerged some 2 billion years ago.

Today, lifelong cave-dwelling organisms, subterranean microbes, deep-sea hydrothermal vents and suchlike may never actually encounter sunlight, but they depend on it nonetheless because they use oxygen generated elsewhere by photosynthesis. Such oxygen is also used by bacteria that exploit chemical transformations – such as those involving sulphur or iron compounds – to 'chemosynthesise' their organic matter. A few anaerobic bacteria can chemosynthesise organic matter using energy from the oxidation of hydrogen by sulphate, carbonate or nitrate instead of oxygen, but their hydrogen is generated by other bacteria from previously photosynthesised organic matter.

Nevertheless, there is no reason in principle why 'mineral' hydrogen, which could be generated chemically from water, might not act as a primary energy source independent of sunlight, supporting an ecology based wholly on chemosynthesis. Such a process, based on the geochemical decomposition of water, has been suggested as a means of supporting deep subterranean microbial life, but the basic chemistry remains speculative. A comparable process could conceivably support life on a warm, wet celestial body, perhaps even on our distant neighbour Europa, one of Jupiter's moons.

John Postgate
Lewes, East Sussex, UK

There are food chains that do not depend upon photosynthesis for the initial input of energy into the system. Chemoautotrophs are microorganisms that obtain energy by oxidising sulphides in inorganic minerals. Hydrothermal vents that pump hot, chemical-laden water onto the ocean floor support flourishing ecosystems far from the influence of sunlight.

These ecosystems derive their energy entirely from the chemicals in the vented water. Nevertheless, the oxygen they utilise to oxidise the sulphides is dissolved in the sea water and is present as a by-product of photosynthesis far above at the depths to which sunlight can penetrate. So even these communities can be said to be indirectly dependent on sunlight and photosynthesis.

Jonathan Wallace
Newcastle upon Tyne, UK

Spreading the love

How does nature prevent incest and therefore inbreeding in animals? Without social conditioning, what happens to discourage (if not prevent) this?

Ann Gilmour
Belfast, UK

In nature, inbreeding is not always undesirable or prevented. Intensive inbreeding can in certain cases purge harmful recessive genes from a population, while ensuring that essential genes are passed on. In a sparse population, it can ensure a mate. In its most extreme form, some plants self-pollinate so intensively that outcrosses are exceptional. Conversely, many plants separate pollen production from stigma ripening to reduce self-pollination, either by wind or by pollinators. Others, such as almonds, are self-sterile, having a sort of immunity to their own pollen.

Some animals are inhibited from incest by clues to shared genes, such as communal odours, although commonly this is incidental to the ejection of adolescents from parental territory, particularly young males when they begin to show cues that identify them as competitors.

While some rodents accept incest, a pregnant female may resorb embryos conceived this way if she smells an unknown male. Analogously, in animals that live in social units, such as lions and some primates, incoming males commonly kill the immature young of their predecessors, or cause pregnant females to abort. This is actually a competitive measure, but it does reduce incestuous reproduction.

Jon Richfield
Somerset West, South Africa

One factor thought to deter inbreeding is the idea of

attractiveness. In a study looking at what makes people seem more attractive to the opposite sex, people were shown two images of the same face. When pictures of the faces were doctored to look slightly more symmetrical, people found them more attractive than the original photographs. When animals inbreed, the chance their offspring is in some way defective, or asymmetrical, is higher. This means that the asymmetrical offspring appear less attractive than their interbred counterparts, and are less likely to find a mate and repeat the habits of their parents or siblings.

In another study, participants were shown pictures of people of the opposite sex who resembled them. In general, the greater the similarity, the weaker the attraction people felt. The likeness that carries through families seems to inhibit incest. If a sibling looks like you, you are less attracted to him or her. So, in theory, even if society was impartial to incestuous relationships, we are genetically hardwired not to find members of our families attractive.

Rob Hayes
Skibbereen, Ireland

Back to the past

Many people suffer some kind of back pain, often crippling. Is it because humans haven't been bipedal for long enough for evolution to have perfected the art of walking upright? And if we are in pain now, what kind of pain must the early bipeds have endured?

Lucy Dodwell
Great Missenden, Buckinghamshire, UK

I would suggest that the problem is the opposite: we are devolving and, as a result, suffering from back pain.

For thousands of years our ancestors walked everywhere, hunting for food. If these hominids were constantly in pain, I assume they would have gone back to the trees to resume their earlier, fruit-based diet. Instead, their modern African equivalents are still walking, sometimes taking days to catch their prey, and until relatively recently, humans used muscle power to grow and harvest crops.

But now we live indoors with a sedentary lifestyle and, consequently, we don't exercise our supporting muscles. Hence, back pain.

John James
Watford, Hertfordshire, UK

Evolution is not in the business of perfecting anything. As R. McNeill Alexander, the comparative anatomist, put it in his book *Bones* (Macmillan, 1994):

> Evolution... starts from an existing design and alters it progressively by a series of small changes over many generations. The final product (for example, an amphibian) may be very different from its ancestor (for example, a fish), but every stage in the evolutionary sequence must be an effective design, capable of holding its own in a competitive world.
>
> As a result, many of the features of a species may not be ideal for its way of life, but may be (more or less) the best that evolution could do, starting from that species' ancestors.

This might lead you to infer that our back structure and upright posture are less than ideal, hence the tendency for so many people to have back problems. Numerous books take this line, implying that we learned to stand upright too quickly, evolutionarily speaking, and have not yet fully adapted to it. This may be true to some extent, but there are facts that support an alternative explanation.

For example, a great number of people never suffer from back pain at all. Does this mean that those who do have a less 'evolved' back? I have heard that argument and, if true, it would seem to condemn those with bad backs to a lifetime of pain.

However, it is possible to reverse the descending spiral of damage and discomfort, returning to a healthy back and a pain-free existence. In my own case, I suffered from a moderate amount of back pain decades ago in my twenties, but through the Alexander technique, I became aware of how many bad habits I had acquired. I had a slumping posture, chronic tension, was habitually bending in my lower back rather than at my hip joints and so on. Once I learned how to carry myself in a more balanced manner throughout all my activities, these pains disappeared. As a bonus, all movement is now easier and freer than previously.

This makes us similar to most other animals, with a system that generally functions quite well. We differ from them in that we are capable of creating furniture and devices which may look nice, but force us into using our spine badly, and have lifestyles that lead us to lose touch with our once natural movements.

It seems clear that the incidence of back problems is increasing in our modern societies as we leave the land and other physical work for a more sedentary life in offices and cities.

I speculate that our ancestors had far fewer back problems than we have now.

David Gorman
Anatomist
Toronto, Ontario, Canada

I suspect that bipedalism has little to do with the back problems we suffer. Over the course of 17 years as a veterinary surgeon, I have diagnosed and treated countless back problems in dogs, cats and even rabbits. According to my knowledge, the frequency and severity of the problems they encounter appear to be much the same as our own.

As far as dogs are concerned, there are undoubtedly cases where selective breeding has predisposed them to slipped discs, and an overenthusiastic use of restraining devices is responsible for a fair proportion of neck problems. But non-pedigree dogs that have not been dragged about by their necks also get backache and neck ache from various causes.

Non-pedigree domestic cats usually experience few or no unhealthy selection pressures from humankind but, though they suffer from far fewer back problems than dogs, these issues still occur.

In summary, quadrupeds can get backache too.

Paul Adkin
Cheltenham, Gloucestershire, UK

The bottom of it

Non-human animals do not use toilet paper, and those that I have observed don't appear to need it. Are there anatomical reasons for this? If so, why is our anatomy not similar to that of the great apes? Has our invention of toilet paper, and whatever methods preceded it, meant that we have lost an anatomical feature that we once had?

Martin Hillman
Edinburgh, UK

Although we share most of our DNA with great apes, there are some striking anatomical differences between ourselves and our nearest relatives, most notably our vertical posture. This enables us to walk tall with our hands free, but it also comes at a price: we experience problems with our back and joints, and the whole business of evacuating our waste is more difficult.

The fundamental problem is that the area used for releasing urine and faeces is compressed between thighs and buttocks, so we are more likely than other animals to foul ourselves.

We also differ from other animals in our response to our waste, which we tend to regard with disgust. This seems to have developed as a result of living together in settlements rather than roaming through the forest, where we could leave our mess behind us. Unlike other primates we can learn when and where it is acceptable to excrete.

Disgust is a sensible response to the threat of pathogens in human waste, and civilisation itself would be impossible without some system of sanitation. We have house-trained ourselves to the point where we are repelled by the very smell of our waste, and thorough cleansing has become a necessity for social reasons as much as for hygiene.

Human ingenuity in this respect has now gone far beyond toilet paper and wet wipes. In Japan there are toilets which will wash and blow-dry your most delicate areas without any effort on your part.

On the other hand, the 16th-century French writer François Rabelais recommended using the softly feathered neck of a live goose for the ultimate in cleanliness and comfort.

Christine Warman
Saltburn-by-the-Sea, North Yorkshire, UK

Although animals seem to have no such inhibitions, the anal area isn't one that humans like to think about. Yet cleanliness around this part of the body is crucial to survival, as it is particularly susceptible to bacterial and parasitic infection.

The anus is a sphincter, a muscular ring, under voluntary control. A useful aspect of its design is that a slight prolapse of the rectum occurs when defecating, minimising contamination of the external area. Also, taking up a squatting position ensures that faeces do not come into contact with any other part of the body. Contamination does occur sometimes, though, and this is where good grooming comes in.

Wild animals, especially carnivores whose faecal matter contains material attractive to pathogens, have evolved to be able to clean themselves. You only have to watch cats 'playing the cello', as it is colloquially called, to see how proficient they are at grooming their rear. Parents will clean their young until they are supple enough to do it themselves. Adult animals will also groom each other, forming social bonds at the same time.

Domesticated animals selectively bred by us are a different case. My dog, for example, cannot clean his hind quarters because he is too short and stocky in the body; we have to check that he is clean after he has defecated.

Similarly, sheep have to be inspected regularly because their body shape prevents them from keeping themselves clean.

Many species, humans included, have adapted their front legs to be hands – helpful for self-grooming. The use of plant material to clean the anal area would have been an evolutionary adaptation. Vegetable matter was substituted with a sponge on a stick in Roman times, and more recently with paper.

Tony Holkham
Boncath, Pembrokeshire, UK

The eyes have it

My 9-year-old daughter has stumped me with a question. She asked why the shape of eyelids differ in different races. She used the example of Japanese versus Caucasian people. Any thoughts?

David Candlish
Chester, Cheshire, UK

The shapes of eyelids, along with other physical features, are the result of evolution – the selecting and deselecting of various traits through time.

A mutation, or random change, in a gene can result in an inheritable physical characteristic that can be beneficial, harmful or neutral. A beneficial mutation may be selected for, so that it spreads from parents to children through a population in a relatively short time in evolutionary terms – as little as 50 generations or 1000 years. A mutation that gives only minor benefit may take longer to become common, perhaps 1000 generations or 20,000 years.

Harmful mutations would be selected against – a euphemistic way of saying that the carrier would die before reaching reproductive age or be less likely to pass on the mutated gene to future generations.

A neutral mutation, which confers no benefit or hindrance, is most interesting. It could survive in the gene pool, neither increasing nor decreasing in prevalence in a population. But if that population moved to a different environment, or the environment changed – for example, because of climate change – then that mutation could be selected for and the gene, with the trait it confers, would become more common.

Some hypothesise that the epicanthic fold that causes the almond-shaped eyes in east and central Asian populations evolved to protect the eye from extreme weather conditions or high levels of ultraviolet light, such as you might get on the

high Mongolian steppes. Whatever the truth of this hypothesis, a population's physical features are no fluke. The genes that are expressed through these features have been tested and selected for over thousands of generations and this continues.

David Muir
Science Department
Portobello High School
Edinburgh, UK

The epicanthic fold of the upper eyelid conceals the inner corner of the eye, called the medial canthus. It is usually accompanied by a lower nose bridge and is most prominent in women and children.

It is thought that the fold helps to protect the eyes from bitterly cold winds, such as those blowing across Siberia, while the narrow nasal passages associated with the lower nose bridge would help warm and moisten inhaled air before it reached the lungs.

This fold also makes it easier to narrow the eyes so that the eyelashes can act like a veil, trapping warmer air near the cornea and reducing the intensity of light reflected off any ice or snow. This may explain why the epicanthic fold is seen in north-east Asia, while migration might explain why it has spread to South-East Asia.

But the epicanthic fold is also observed in disparate groups across the globe. In the bushmen of southern Africa and indigenous Americans, it may have afforded protection against dust storms: the fold shields the lacrimal glands, which produce tear fluid to lubricate the eye and cleanse it of dust, and nasal hairs in narrower nostrils may trap more dust.

The trait can also be seen among Scandinavians, Sami, Poles and Germans, as well as Irish and British people, which may be the result of migration.

Mike Follows
Sutton Coldfield, West Midlands, UK

I was surprised that neither of the previous answers mention sexual selection, Darwin's other great insight into evolution.

A memorable passage in Jared Diamond's *Guns, Germs and Steel* quotes a man from the New Guinea highlands on the great beauty of the local women, with their tight curly black hair, their eyes very close together, and their noses spread right across the face. He contrasts them to ugly western women, with their hair like dead grass, their eyes wide apart like a pig's, and their axe-blade noses.

These are all features for which it is hard to assign any evolutionary advantages, but which are easily accounted for in terms of sexual selection. It may be that the epicanthic fold confers no evolutionary advantage, but is considered more beautiful in some cultures, and so predominates.

Stephen Thomson
Ashfield, New South Wales, Australia

Dead end?

As we further develop technologies that enable us to survive changes in our environment, does this mean that we will stop evolving?

Peter Hoare
Kings Lynn, Norfolk, UK

This question of whether humans are still evolving has been the subject of much interesting debate of late. Medical advances mean that our physical and reproductive health are no longer the major determinants of our ability to pass on genes. This means that natural environmental factors will have less influence on which people pass on genes than in our distant past. However, the process of evolution will still be at work.

It has been suggested that within many cultures women now have greater freedom and choice in partners, so arguably traits that women in our culture find attractive are more likely to be passed to the next generation. One might think that in our modern technological world, genes for high intelligence would be favoured. However, success in modern societies seems to cause those successful people to have fewer children, not more. So perversely, genes for intelligence may be being selectively bred out of the population.

Alternatively, the human species could conceivably evolve into two separate genetic pools: an intelligent and affluent pool with low reproductive rates and longer lifespans because they can afford the latest medical treatment; and a larger pool of the lower intelligence, poorer, exploited class that has a high reproductive rate and lower lifespan. However, there is probably too much crossover of members between the two groups for this to happen, and it is also questionable whether such a society would be stable long enough to form two separate species. Long live the revolution!

Simon Iveson
University of Newcastle
New South Wales, Australia

Human beings will not stop evolving because of technology developed to enable us to survive changes in our environment. Consider that technology has dramatically reduced infant and maternal mortality in the developed world. Presumably, babies with larger heads are more likely to survive, along with their mothers, due to procedures such as caesarean sections. This should lead to an increase in average head size.

Similarly, technology enables people with poor eyesight, and other conditions that once would have been fatal, to survive and reproduce. Couples can overcome infertility through assisted reproductive technology, thus spreading their genes for low fertility more than would have been possible in ancestral environments.

These are just some examples of how technology alters the direction of evolution, although perhaps not in the way the questioner would consider desirable.

Ellen Spertus
San Francisco, California, US

The advance of technology does not mean the end of human evolution. Our species is evolving faster than ever. Evolution is caused by natural selection acting on inherited variation. With our population at 7 billion, we are more genetically diverse than ever; and selection pressure has shifted rather than ended.

Civilisation exerts powerful selective pressure against lactose intolerance, coeliac disease, dyslexia, innumeracy, immune deficiency and – in this materialistic age – a lack of sales resistance.

Our far-future descendants will, from a very young age, be highly literate, numerate, have strong immune systems, and be resistant to stress; and they will also possess a superhuman sense of humour. They will need those traits to survive long enough to reproduce.

Nathaniel Hellerstein
San Francisco, California, US

7 Biology

Body heat

What mechanism in mammals provides the temperature reference point that enables humans and other animals to have such accurate control of body temperature?

Mike Tongue
Newark, Nottinghamshire, UK

Body temperature in mammals is controlled by a region of the brain called the hypothalamus. This monitors the temperature of blood flowing from the heart through the brain, as a measure of the body's core temperature – the temperature of the heart, liver and so on. If the hypothalamus detects a move away from the normal 37 °C in humans, it coordinates a series of responses to lower or raise body temperature. If a rise is detected, it triggers sweating and dilation of the arterioles running close to the skin in order to lose heat. If the core temperature is too low, it encourages an increase in metabolic rate, constriction of the arterioles near the skin and erection of hairs on the skin to conserve heat. If body temperature continues to fall, it triggers shivering to generate heat.

Katherine Gourd
London, UK

The basic temperature control mechanism is a 'thermostat' in the hypothalamus whose neurons fire more rapidly when they get warmer and more slowly when the temperature drops. These cells use hormones to signal changes in the basal metabolic rate, send nerve signals to initiate shivering, sweating or panting, or redistribute blood flow to and from the limbs and skin to lose or conserve heat.

The questioner suggests that mammals set a precise temperature: they do not. For example, our temperature rises when we are fighting infection, which helps to kill or disable it. In fact, syphilis used to be cured by infecting patients with malaria. Several fever cycles would kill the syphilis, and the doctor could then administer quinine to treat the malaria.

During exercise, our body temperature is allowed to rise to between 39 and 40 °C. You know the body 'allows' this to happen because you do not start to sweat until your body temperature reaches the new set point appropriate to the increased level of activity.

Other animals can also vary their body temperature. Under normal conditions, desert mammals maintain a temperature of around 37 °C. But when deprived of water, some species allow their bodies to warm up considerably to 39 °C or more during the day. By so doing, they save a lot of water that would otherwise have been used for evaporative cooling. At night, when the temperature falls dramatically, they allow their body temperature to fall well below the normal 37 °C set point to as low as 35 °C.

Some desert antelope, such as the east African oryx, let their body temperature rise as high as 45 °C, which would be hot enough to thoroughly addle their brains. However, their panting preferentially cools the arterial blood feeding their brains, allowing them to maintain a brain temperature of around 39 °C while the bulk of their body is about 5 °C warmer. This also saves water that would otherwise be used for cooling.

Most animals allow their extremities to cool considerably below their core temperature. This is done partly by reducing

the rate of blood flow to their limbs and tail, and partly by directing the cool venous blood returning to their heart into a set of veins running close to the arterial supply. This heat exchange cools the arterial blood flowing to the limbs, and warms the returning venous blood, thus retaining body heat and saving metabolic energy.

Peter Bursztyn
Barrie, Ontario, Canada

Gasping for nitrogen

Life evolves to adapt to its environment. So how come life most frequently requires oxygen for survival rather than the more abundant nitrogen?

Karsten Stier
Port Fairy, Victoria, Australia

Organisms need a source of energy to drive the processes of life. Respiration creates that energy. Molecular oxygen is a very powerful oxidising agent, reacting with substances such as glucose and releasing energy in the process. Molecular nitrogen, on the other hand, is relatively unreactive at room temperature and is therefore pretty useless for providing that energy.

Simon Iveson
Negeri Yogyakarta University, Indonesia

While there is more than enough oxygen to sustain respiration, plain nitrogen is too unreactive to be of much use. What's more surprising is that most life has come to rely on another gas that makes up only 0.03 per cent of the atmosphere, namely carbon dioxide. Photosynthesis depends on it.

So life evolves not only to fit the inorganic environment, but to take advantage of inalienable chemical processes.

Life does not entirely shun nitrogen, however. Some bacteria take it from the air to convert into the nitrogenous substances that plants need to build proteins. The leguminous plants have even evolved roots that offer these bacteria ideal habitation, although most plants can get nitrogenous nutrition from other indirect sources, such as ammonia.

Paul Schifferes
London, UK

Life uses oxygen instead of nitrogen for the same reason we burn gasoline in cars rather than the more abundant water: it's all about oxidation.

Animals and plants respire through breaking down sugars by the addition of oxygen to create carbon dioxide, water and energy. This is oxidation.

Unlike molecular oxygen, nitrogen gas is exceptionally stable and requires the input of energy to break it down into usable forms. So although nitrogen is a vital constituent of proteins and DNA in all living organisms, very few of them can utilise nitrogen gas as a source.

Nelson D. Sherry
Oregon, US

If nitrogen were to become involved in our metabolic processes it would need to be fixed by the specialised and energy-dependent processes found in bacteria associated with leguminous plants, or by reactions induced by lightning. It would then make its way to us via these plants, ready-packaged as proteins and other nutrients. The reactivity of oxygen, however, means we are able to use it in its gaseous form.

Steven Tait
University of Edinburgh, UK

Reflection reaction

When an animal looks in a mirror, does it realise that it is looking at itself? Which, if any, animals successfully make this connection?

David Vincent
Southampton, Hampshire, UK

Experiments in primate self-recognition were done in the 1970s by Gordon Gallup. He anaesthetised chimpanzees, then marked their faces with blobs of non-toxic paint. The chimps were given mirrors to look in when they woke up. They saw the blob in the mirror, touched it and cleaned it off. Apparently orang-utans can also do this, but this behaviour has so far not been recorded in gorillas, monkeys or any other animal apart from humans.

Penny Hawkins
Horsham, West Sussex, UK

This is still not a completely solved issue. Following Gallup's experiments, many researchers found behavioural evidence for self-recognition in apes (chimps, bonobos, gorillas and orang-utans) and maybe in dolphins.

Besides the experiments that involved marking faces, others were devised in which the animal was urged to do something with its hand, which was hidden behind a screen, while being shown a television picture of its hand in reverse. Chimps learned to manage the task by looking at the screen and moving their hand in a trial-and-error strategy guided by visual feedback. Other tasks investigated whether the animal recognised itself in a videotape image, a photograph, or even its own shadow.

Of course, all these experiments are behavioural tasks which cannot access the mind states of these animals directly and prove if they really do recognise themselves. Some researchers are therefore still sceptical.

Macaques and other less intelligent primates have not yet displayed the self-recognition abilities of apes and humans. And apes that grow up with very limited social contact seem to fail the task too. Furthermore, developmental psychologists found evidence that self-recognition correlates with empathy. It is possible that both abilities emerge together.

For a good review of the whole subject I can recommend *Self-Awareness in Animals and Humans* edited by Sue Taylor Parker, Robert W. Mitchell and Maria L. Boccia (Cambridge University Press, 1994).

Incidentally, human babies normally learn to recognise themselves in a mirror at between 12 and 20 months. Once I sat with a very young girl in front of a mirror, asking her what she could see. 'That's me,' she told me. I repeatedly insisted that it couldn't be her, because she was next to me (I touched her) and not there (I touched the mirror). After a while she said: 'That's a picture of me.'

Rudy Vaas
Bietigheim-Bissingen, Germany

A bird regards the reflection as another individual of its own kind and does not realise it is looking at itself.

Several species of territory-defending Australian birds will spend lengthy periods attacking their reflection in, for example, the wing mirror of a vehicle, or a flat chrome hubcap. The magpie lark (*Grallina cyanoleuca*) is a notorious offender. It is a practice which doesn't endear the bird to the motorist, for the mirrored surface becomes smeared with oil from the bird's feathers.

The English ornithologist John Gould regarded Australia's superb lyrebird (*Menura superba*) as the shyest and most difficult to approach of all birds. But male lyrebirds are

strongly territorial in the breeding season and in the course of research on the species, CSIRO, Australia's national research organisation, showed them capable of being captured by using a mirror at the back of the trap.

Syd Curtis
By email, no address supplied

Ant party

The waggle dance of bees has been well reported, but how do ants know to gather in large numbers when one of them, presumably, has discovered a tasty morsel? Does it report back to the nest and organise a foraging party?

David Prichard
Bluff Point, Western Australia

Ants forage randomly. On their way back to the nest with food they leave a pheromone trail that other ants will randomly encounter and identify as 'pointing' to a food source, according to the concentration of pheromone at each point.

The more food, the more pheromone, and the more ants will pick up the trail. It's a fascinating and simple system they have used for millions of years. Pheromones evaporate over time, preventing areas from becoming saturated and confusing. Many studies have modelled ant foraging behaviour; some are impenetrable, but others can make very interesting reading.

Ants are generally unwelcome in the house, and the best way to prevent them coming in is to follow a homeward-bound individual to its point of entry and dowse about a square metre around the area with boiling water, followed by a small amount of anti-ant powder.

I carried out a comprehensive study of ants while at college and, although insects in general are fascinating, I think ants are the most interesting of all – especially in the way they nod to each other when passing on a route, just like truck drivers do.

Tony Holkham
Boncath, Pembrokeshire, UK

Ants communicate with one another through pheromone release. The pathfinder forager ant leaves pheromone trails from the spot where the tasty morsel is residing all the way back to the colony – hence it is able to organise a foraging party.

Other ants follow the pheromone trails and also reinforce it, but once all the food is consumed, they stop leaving pheromone trails and disappear. One in five foragers are pathfinder ants that can detect previously marked pheromone paths for up to 48 hours

By email, no name or address supplied

Lack of splat

How can flies fly at speed into a pane of glass and seemingly remain uninjured?

Paolo Lanzarotti
Pesaro, Italy

The anatomy of flies is springy, but the main reasons they survive are matters of scale. Although flies fly at huge speeds relative to their size, their actual air speed seldom attains 3 metres per second, and is usually nearer 1 metre per second. Given that kinetic energy is 0.5 multiplied by mass multiplied

by velocity-squared, the amount of energy per milligram of fly is tiny. In hitting a glass pane, it will barely bend its own bouncy bristles.

Hitting a moving sheet of glass is a different matter. At modest highway speeds, a car windscreen travels more than 30 times faster than a dawdling fly. The resulting squared velocity inserted into the equation above means that the imparted energy is perhaps 1000 times as great. The fly is of negligible mass relative to the car, so essentially it absorbs all that energy and the windscreen bears a red streak of the fly's optic pigments. If you suck a fly into a vacuum cleaner you may find a similar souvenir inside.

Jon Richfield
Somerset West, South Africa

While that housefly on a collision course may appear to be zipping along, this is an illusion caused by its small size: its top speed is a modest 2 metres per second. For the blowfly it is 2.5 metres per second. Weighing 12 and 50 milligrams, respectively, the force with which the two hit the pane will be at least six orders of magnitude less than that of a 70-kilogram human walking into patio doors at 5 kilometres per hour (or 1.4 metres per second).

Kinetic energy imparted on impact is proportional to mass and velocity squared, so we find that the human headbutts the glass with an energy between 400,000 and 3,000,000 times as much as the flies' bodies. To impart the same energy as a walking human, let alone a runner at full pelt, the insects would have to be doing between roughly 6000 and 12,000 kilometres per hour – or Mach 5 and 10 – respectively.

Most human injuries from impacts with glass comprise bruising of superficial tissues or, if the force is sufficient to break the glass, cuts from shards and sharp edges. Hard impacts can cause fractures or concussion.

Flies, in contrast, have a stiff, tough outer skeleton of chitin in a protein-rich matrix. The chitin between the body segments and joints of the appendages allows the shock of impact to translate to movement of appendages about the joints. The fly is its own shock absorber.

Chitin is akin in its chemical and mechanical properties to the keratin of human fingernails, and the lack of injury to a fly upon hitting the pane might be likened to our ability to flick a paper pellet without harm to our fingernail or the quick beneath it.

Len Winokur
Leeds, UK

Big fish in a ...

Is it true that goldfish (or other captive fish) grow in proportion to the size of the tank they live in? I have a small goldfish in a tank at home but have seen much larger goldfish in ponds and believed they were a different species. But a friend insists they are the same, with the size difference being caused only by the volume of the body of water in which they live. Is my friend right? And if so why?

Barry Dwyer
Truro, Cornwall, UK

It depends. Some Koi, a related but different species, are easily confused with large goldfish. Real goldfish have been bred and cross-bred for more than 1000 years, and are now regarded as a single species. But there are various breeds that differ in size, even allowing for pond volume.

Water supply is seldom a constraint on marine fish, but ponds and streams are another matter. Freshwater fish are

generally adapted to grow according to available resources. I successfully raised baby guppies in a wine glass and they grew apparently healthy and happy, but were less than half the normal size and produced very small litters. Much the same applies to other freshwater fish and even tadpoles, and multiple factors seem to control their growth.

Temperature, depth and mass of water seem to be important, as do famine and crowding. Apparently, a common control factor is the secretion of certain protein molecules into the water by the fish. These stunt the growth of young fish, and affect the smallest more rigorously than the larger ones. So commercial fish breeders need to be lavish with the turnover of fresh water if they are not to be stuck with unsellable runts.

Antony David
London, UK

What determines the ultimate size of goldfish will include genetics and the availability of resources such as food. But there are social factors at work too, and these are of particular interest to ecologists, fisheries biologists and fish farmers.

Goldfish are schooling fish, meaning that they naturally form loosely coordinated groups that spend variable amounts of time together. Shoaling fish such as herring, by contrast, form tightly bound groups that swim and forage together all the time.

Within the groups, bigger goldfish are dominant, and they occupy the best microhabitats and take advantage of the best sources of food. Such fish are able to grow faster and avoid predators for longer, and this in turn increases their dominance.

While all fish grow throughout their lives, they normally grow fastest when young, and the rate tends towards zero as they age, so if a goldfish does not become dominant in its first year, it is unlikely ever to reach its genetically possible maximum size, no matter how long it lives or how well it feeds in later years.

This explains the common situation in which a pond is occupied by one or two big goldfish and lots of smaller ones. Even if the large fish are removed, the remaining ones will never get much larger.

However, the size of aquarium and pond fish is usually more to do with environmental factors than social ones. Indeed, the mbuna cichlids of Lake Malawi, which are some of the most intensely social fish, commonly get bigger in aquaria than in the wild because they are given a much richer source of nutrients (in the form of fish food) than they would get in the wild (algae and, to a lesser degree, micro-invertebrates).

There's no reason goldfish should become stunted in ponds if water quality is good and food supply rich and varied. But because garden ponds are often filtered (to remove fish waste) and almost always receive few water changes, which would dilute the end products of biological decay and filtration, water quality in ponds may not be as good as the pond keeper thinks. This, more than anything, explains why goldfish in ponds are smaller than expected.

That we so often see stunted goldfish in ponds, aquaria and bowls says less about their ability to grow to the size of the container and more about their ability to hold onto life where other fish would simply die.

Neale Monks
Peterborough, Cambridgeshire, UK

The author writes about fish health for Practical Fishkeeping *magazine and is the editor of* Brackish-Water Fishes: An Aquarist's Guide to Identification, Care and Husbandry *(TFH Publications, 2006) – Ed.*

Perchance to tweet

Birds sing at dawn and tend to go quiet after dusk. Presumably during darkness they sleep. But the further from the equator one lives, the greater the disparities between the hours of darkness and daylight. So in summer, in temperate or polar areas, do birds not get enough sleep? Or do they get too much in winter? How do they deal with the changes?

Peter Chandler
Rochester, Kent, UK

Birds do sleep mainly at night, but only in very short snatches. They will also sleep during the day provided they feel secure enough, so the length of daylight makes very little difference to them.

They sing for several reasons, including to attract a mate and to keep in touch with the flock. But the principal reason is territorial. We hear them mostly at dawn because it is quiet and therefore the best opportunity to re-establish territories. Most sing during the day, too, but this tends to be drowned out by a cacophony of traffic, music and the ubiquitous power tool.

Some birds do sing during darkness: robins, thrushes and nightingales are common examples. But other birds such as rooks and magpies will chatter away in the dark, too, and I have heard sparrows chirping very quietly in the hedge at night, particularly when sitting on eggs.

Street lighting can confuse birds. Last year when the lights in our village were on all night, birds sang in the middle of the night.

Tony Holkham
Boncath, Pembrokeshire, UK

Birds cannot risk the long, deep sleep to which we are

accustomed. They need to be alert to danger, so they sleep in short bursts. Moreover, many species have long migration flights, which represent an even bigger assault on their sleep than seasonal changes in the hours of darkness. It is said that the common swift spends virtually its whole life in the air, landing only to lay its eggs and raise its chicks.

Many birds migrate at night to avoid predators and overheating. Individual legs of a migration may exceed 24 hours, especially when passing over an ocean or other geographical barrier. Instead of allowing the whole brain to sleep, birds can let each hemisphere of the brain and its associated eye take turns to rest, in a process called unilateral eye closure. This is the biological equivalent of autopilot – the active half of the brain can still alert them to danger.

Between flights, Swainson's thrushes and other migrating birds catch up with missed sleep by taking significantly more 'micro naps', lasting just 9 seconds each, and drifting into an intermediate sleep-like state referred to as drowsiness, when they keep a lookout through drooping eyelids. They supplement this with unilateral eye closure. Although not as rejuvenating as normal sleep, this has the advantage of decreasing the risk of being caught and eaten.

Amba Sanghera
Mtarfa, Malta

Buzz off

I have the back door open and a bee has entered my kitchen and is buzzing around. Is the best course of action to leave the door open and hope it goes out, or is there a greater chance of more coming in?

Simon O'Callaghan
London, UK

It depends on the bee. Honeybees and bumblebees often end up indoors more by accident than intentionally. They seldom find the open window or door to leave again and will tend to go towards the light. This often results in them banging against the window for ages. It is simple enough to pick them up by the scruff of the neck and put them outside. If you don't fancy that, an inverted cup with a stiff piece of card will suffice.

Wasps are a different matter. They will buzz around looking for something to eat. Sometimes they, too, are attracted to light and will go to a window, but they will often find their own way out. Again, if in doubt remove them yourself – but don't get stung. Lights on inside a room will confuse both insects, and they will buzz around them.

As a beekeeper, I find bees coming indoors when I'm extracting honey. They will even stay when it gets dark, but wasps leave by the way they came in when light levels drop. A single bee entering is usually not troublesome. However, there is a slim chance that a swarm is roosting and a scout has come in to check out the accommodation.

Gerald Legg
Hurstpierpoint, West Sussex, UK

I have found that pointing to the door and saying 'out' firmly will lead to the bee leaving almost immediately. It doesn't work with wasps or flies.

Rod Newbery
Cambridge, UK

This question reminds me of the (possibly apocryphal) report about a caller to a US phone-in radio programme who said they had a skunk in the basement and asked what was the easiest way to get rid of it. Within minutes another caller suggested leaving the exterior door open and laying a trail of crumbs from the centre of the basement to the door. Half an hour later the original caller phoned back and said there were now two skunks in the basement.

John Darlington
Annaduff Glebe, Ireland

Four legs good

Because there are no roads in the Himalayas, goods have to be carried either on porters' backs or by pony trains. The tracks are steep, often consisting of uneven steps. The ponies need to keep looking at the track immediately ahead in order to position their front hooves. But how do they know where to place their hind hooves?

Bob Miles
Sidmouth, Devon, UK

This is a vital skill for large quadrupeds for whom tripping is dangerous, such as horses and elephants, but also for animals that must stalk quietly when hunting, such as wildcats. These animals automatically guide the hind leg over obstacles the forefoot has already negotiated. Surreptitiously remove an obstacle once a cat's forefoot has passed over, and the hind leg unnecessarily follows a trajectory over the absent obstacle; but introduce an obstacle after the foreleg has passed and the hind leg bumps into it.

The characteristic walking gait of tetrapods plants the hind foot just behind the forefoot. Usually footing will be secure there, because it has already been tested. Similarly, chameleons negotiate a wide gap by anchoring their tails and hind legs, reaching out with their forelegs, then swinging the hind legs forward to where the forelegs have gripped. Only then does the tail let go.

Galloping is an emergency gait and accordingly riskier, often with rear feet landing well ahead of forefeet, but even then most large animals are surprisingly good at directing their feet. Small prey animals such as some rodents have a contrasting strategy. They systematically practise panic retreats along established paths so that they know the location of every pebble. This means that in emergencies when noise is no consideration they can flee blindly at top speed.

Jon Richfield
Somerset West, South Africa

A horse has a number of gaits, as humans have when walking and running. In the walk and the trot, the rear foot falls at almost exactly the same place the front foot on the same side was positioned. Thus a horse trained on difficult terrain will select a foothold for the front foot, and be able to aim its back foot at that spot without looking.

It relies on proprioception, the ability by which humans and other animals can unconsciously sense the position of joints, and via the tension of our muscles put a finger on our nose, or swap an object between our hands, without looking.

However, for horses this does not work at faster gaits such as cantering and galloping, and a horse won't go this fast on difficult terrain, nor would its rider want it to!

You can see different gaits at bit.ly/18nBy2k.

John Davies
Lancaster, UK

Many years ago, I ran down a rocky mountain path and was very aware of the need to place my feet strategically to avoid injuring myself. I noticed that I wasn't looking at my feet at all – I was looking a pace or two ahead. It seemed that, having planned a step, my brain planned future steps before the first one had even been executed.

I wondered whether similar planning happened in easier walking situations. Back in town, I discovered that, as I walked along, I could pick a crack in the pavement about five steps ahead and instinctively predict with a high level of success which foot would step on the crack, or, if my legs were going to straddle the crack, which would be the leading foot. And this was happening having never tried the task before.

This all suggests that walking involves much more forward planning than we are conscious of.

Ben Craven
Menstrie, Clackmannanshire, UK

Big blue

Why do blue whales need to be so big when they only eat plankton?

Peter Hammond (aged 7)
Toronto, Ontario, Canada

Blue whales need to be so big precisely because they eat tiny food, specifically krill, small crustaceans that feed on plankton.

Krill defend themselves against smaller predators by forming very dense shoals that confuse attackers. Blue whales get around this by swimming very fast at the shoal with their mouth wide open, often from below, and engulfing as much as they can. Their pleated throat can expand enormously to take in as much of a shoal as possible. The whales then strain the krill by forcing the water out through the baleen plates in their mouth.

Because each attack uses up a lot of energy, this feeding

strategy – called lunge feeding – only works if the animal can take in enough krill to more than repay the cost, so the whale needs to be huge to use it effectively.

Peter Wright
Polegate, East Sussex, UK

One struggle for mammals that live in a cold ocean is to maintain their body temperature. Besides having a thick layer of blubber as insulation, another strategy is to be as large as possible, because this minimises their surface area-to-mass ratio. In other words, there is less surface area through which to lose heat per unit of body weight. Young whale calves don't have this benefit, which is one reason why many whale species migrate to warmer waters to give birth. The calves then have the chance to fatten up before they venture to colder waters.

The upper limit on body size is determined by the skeletal structure needed to support the body. For land mammals this is a more serious limit because they don't have the buoyancy of the water to help support their weight. That is why land mammals can't grow as large as whales.

Simon Iveson
University of Newcastle
New South Wales, Australia

Long-haul racers

Can any land-based animal outperform an elite human marathon runner over the full distance? If so, which ones?

Michael McCullough
London, UK

Several land animals could beat a person over a marathon, including the husky, the camel, the pronghorn (a creature similar to the antelope) and the ostrich.

Accurate or not, the story of Pheidippides is the inspiration for the modern-day marathon. He is supposed to have covered 240 kilometres over two days in order to reach Sparta and summon aid when the Persians landed at Marathon in Greece. Later, he ran 40 kilometres from Marathon to Athens and used his final breath to announce Greek victory.

Some might argue that sending a messenger on horseback would have been more sensible. After all, a human has only ever beaten a horse twice in the 33 times that the Man versus Horse Marathon has been run in the Welsh town of Llanwrtyd Wells. However, this race is run over 22 miles (35 kilometres) and not the 26 miles and 385 yards (just over 42 kilometres) of the full Olympic marathon distance.

It is likely that long-distance running echoes a hunting strategy our ancestors developed more than 2 million years ago. Our prowess over distance relies on our ability to avoid overheating, accomplished mainly through sweating and being hairless. Cursorial hunters – those that are slower over short distances but have greater endurance – like humans simply have to run faster than the slowest gallop of a prey animal until it collapses with heatstroke.

Horses are significantly better than people over about half the marathon distance, which is why they were used for the Pony Express mail service in the US before the telegraph was introduced in 1862. Horses were ridden quickly for about an hour between stations, an average distance of 24 kilometres, and the riders would change horse at each station.

Perhaps the Greeks used a runner and not a rider because a horse would not have coped with the heat of the late Greek summer and the mountainous terrain.

During the annual Iditarod trail dog sled race held in

Alaska, the dogs pull the sled at around 24 kilometres per hour for up to 6 hours at a time.

At these speeds, if they were running a marathon, Alaskan huskies would cross the finish line in less than an hour and a half – at least half an hour faster than the human world record.

The pronghorn has evolved to outrun the grey wolf and can sustain 48 kilometres per hour for about an hour. The ostrich has long legs – composed mainly of tendons – that work almost like pogo sticks, storing and releasing elastic potential energy. The pronghorn and ostrich could probably run a marathon in 45 minutes.

Mike Follows
Willenhall, West Midlands, UK

A previous answer pointed out that the pronghorn – a creature similar to an antelope – could run a marathon in well under an hour. The answer also attributed the animal's speed to the fact that it 'has evolved to outrun the grey wolf'.

Maybe, but why is it that the pronghorn very easily outpaces wolves over long stretches, as well as the shorter distances required to escape the predator? A pronghorn's top speed is around 90 kilometres an hour – making it second only to the cheetah – and it may be able to sustain speeds as high as 60 kilometres an hour for about 6 kilometres.

In 1996, biologist John Byers proposed that the pronghorn's great speed is an adaptation to predators that are now long extinct. These predators may have included false cheetahs and endurance runners like the long-legged hyena. It is an appealing idea, although near impossible to confirm.

Michael Le Page
London, UK

Silk stockings

We share our house with a few spiders. This morning, one with a small body and long spindly legs trapped another squatter, darker spider in its web. How can this happen? Since spiders are adept at walking along strands of silk, why should one get trapped in another's web?

Jim Ainsworth
Leominster, Herefordshire, UK

This predatory spider is most probably of the *Pholcus phalangioides* species that was first described by entomologist Johann Kaspar Füssli and is also known as the cellar or skull spider.

It cannot survive a cold winter outdoors in the UK, but it has become quite common in heated houses there. It is a specialist at feeding on other spiders.

I once accidentally dislodged another species of spider, which fell into the clutches of a cellar spider. It was promptly dispatched, even though its body was somewhat larger.

The intruder was quickly pounced on and transformed into a silk parcel in an instant. The cellar spider clearly had the edge in speed and dexterity, which is all the more remarkable for an insect that has such a clumsy appearance.

Contrary to what you might suppose, spiders are not immune from being trapped in their own webs. They avoid getting stuck by leaving some strands of silk untreated with glue, which allows them to deftly pick their way around without becoming entangled.

Blundering into a strange web, the intruder spider in the question didn't know which strands were untreated. It also wandered into the domain of a spider that sees other spiders as a potential meal, rather than a threat.

Terence Hollingworth
Blagnac, France

Spider species can only be accurately identified by microscopic examination, so my explanation is based on the assumption that the spider in the question with the spindly legs is *Pholcus phalangioides*.

This species, once confined to regions nearer the equator, is becoming much more common in the UK as a result of climate warming. It is known to prey on the common house spider, *Tegenaria domestica*.

Pholcus phalangioides can move very quickly, much quicker than a *Tegenaria domestica*. If it notices a nearby threat, such as a finger, it will gyrate so rapidly that it becomes a grey blur. It also has a particularly potent venom, though one which is not so painful to humans.

I would suggest that what your questioner observed is a *Pholcus phalangioides* that had immobilised its prey with its powerful venom and then wrapped it up in silk, ready to consume. The victim's struggling gives the erroneous impression that it has become tangled in the web by its own efforts.

Paul Dunford
Cumbria, UK

Well spotted

What causes freckles? And why do some people have them while others don't?

Maeve Halloran
Dublin, Ireland

A freckle corresponds to a higher concentration of the pigment melanin, and is most obvious when it contrasts with fair-coloured skin. Freckles are associated with variants of the gene on chromosome 16 for the melanocortin-1 receptor

(MC1R), which are also responsible for red or ginger hair. This probably explains why there is a correlation between freckles and red hair.

Melanocytes in the skin produce melanin and package it into organelles called melanosomes. These are passed into overlying keratinocytes, the cells that form the outer barrier of our skin, where they release their payload of melanin. Those born with darker skin have larger melanocytes, which lead to more melanin in the outer skin cells. Freckles are also associated with bigger melanocytes.

Freckles are triggered by exposure to sunlight. UVB radiation activates melanocytes to increase melanin production, which can cause freckles to darken, increasing effectiveness as a sunscreen. The person tans relatively quickly where they have freckles, but the skin between is still prone to burning.

Red hair and freckles occur most frequently in people with northern or western European ancestry. For example, 13 per cent of Scots are redheads and about 40 per cent of them carry the red-hair gene.

Fair skin and freckles might bestow an evolutionary advantage to those living at high latitudes, where it is colder and the intensity of sunlight is lower. It is suggested that a paler complexion reduces heat loss through radiation, though clothing would surely be more effective at retaining heat.

The lighter skin pigmentation between freckles also leads to greater absorption of sunlight and higher production of vitamin D, reducing the incidence of rickets in northern latitudes.

Mike Follows
Willenhall, West Midlands, UK

No pips to squeak

I have a mandarin tree that produces lots of fruit. Last year some of its fruit had pips but the rest contained none. This year I couldn't find pips in any of the fruit. What causes these pips to form, or not?

Malcolm Frost
By email, no address supplied

The short answer is plant hormones. The growth of each part of a plant depends on sensitive and complex feedback that controls several different kinds of hormones. In the case of fruit, structured tissues grow quickly and then die in a certain manner. Anything that affects the controls on this process can make it go wrong, whether the result of genetic changes or damage by microbes or insect pests.

For example, fruit often falls prematurely because something is inhibiting seed formation – developing seeds being the source of hormones that stop the plant dropping the fruit. It makes sense for plants to let these fruits fall rather than waste resources on something that will never produce seedlings. (When ripe fruits fall, they do so in response to different hormones.)

Surprisingly, spraying growing fruit with the hormones that prevent it dropping can, in some plants, yield seedless ripe fruit. This technique can be used to produce seedless melons and peppers, for example. Mutations prized by farmers allow other sterile fruit to ripen, such as navel oranges and seedless grapes.

The tree in the question patently has the genes for producing enough of the right hormones to grow fruit with fewer seeds, or perhaps sterile seeds too small to notice. This is convenient until you wish to propagate the tree from seed. Good luck if you do.

Jon Richfield
Somerset West, South Africa

Hot in the hay

I have always assumed that the belief that haystacks can burst into flames spontaneously was a convenient myth to cover for careless farm workers having a crafty cigarette break while forgetting their surroundings, but a friend insists that it can happen. Surely the only way hay can warm up significantly is if it is wet and bacteria begin to heat the stack as part of the process of biodegradation. Even so, I'd be amazed if this could generate temperatures hotter than about 40 °C. So how else could ignition take place?

Antony Wheatley
Malaga, Spain

Cut hay may look dry, but it is considered inert only once its moisture content falls below about 15 per cent. When moisture levels are above 30 per cent, the tissues continue to respire, generating heat and water, which emerges as vapour through the leaf pores. Within the confines of the bale, the water condenses and spreads by capillary action, promoting bacterial and fungal growth, which adds respiring biomass.

In recently harvested hay, the result can be a single temperature peak of up to 60 °C between five and seven days from the day the reaction begins. This is self-limiting because the temperature kills most microorganisms and drives off the moisture. Sometimes a few weeks of warming cycles can follow as colonies of fungi wax and wane, but the successive temperature peaks likewise dwindle and the bale cools to match its surroundings.

However, if the hay has just been harvested, and the weather is humid, say, or wet with rain or dew, then a stack may sustain these warm conditions long enough to promote heat-loving microbes. These kick in at around 45 °C and die by 80 °C. They are not in themselves dangerous but as they raise the temperatures, they can trigger exothermic

chemical reactions that accelerate with rising temperature. As chemistry takes over from biology, temperatures can rise to a blistering 280 °C. Deep inside the bale this may stop at charring, but when the temperature rises above 231 °C the hay can auto-ignite on contact with air – it does not need a spark or a flame. The sequence varies with bale shape and size, porosity to air, and whether it is stacked or confined (in a barn, for example).

A dry outdoor bale of hay (which is grass and mingled herbs) or straw (which is the cereal stalks after the grain has been removed) could also conceivably catch fire from the sun's heat focusing on a discarded, half-empty glass bottle. I recall reading a news item about a house fire caused by a bowl of water on a windowsill. Forensic science would show such a hay fire to have started on the outside of the stack. Spontaneous fires, on the other hand, originate internally. Farmers know to monitor their stacks regularly.

Similar warming can occur in domestic compost heaps, which makes them popular egg-laying sites for grass snakes. If the eggs aren't actually poached, a two-headed snake often emerges, reflecting an unusual sequencing of critical biological processes. A fuller explanation of the twin-headed snakes can be found in Mark S. Blumberg's book, *Freaks of Nature* (Oxford University Press, 2009).

Len Winokur
Leeds, UK

I grew up on a farm in Suffolk and we used to make a considerable amount of hay each year. Hay is nothing more than air-dried grass. It is made during the late spring and early summer, and before it is mowed the grass is often quite thick and the ground can remain very wet even after long periods without rain. The mower leaves the hay in swathes, so that it starts to dry and allows the ground in between to dry as well.

The drying grass is then turned a number of times using a hay bob, and when it is judged ready, or nearly ready but rain is forecast, the hay is rowed up and baled.

Baling hay, especially with a traditional baler, is an art. When you start baling a field it can often be mid-morning and the dew may only just have lifted. As the day wears on the hay gets drier and so the bale density has to be continually adjusted to keep the bales tight. It is this tightness that can cause issues with combustion.

Hay is an incredibly good insulator and the denser, and hence wetter, a bale is, the less air gets into it. Anybody who spends time in a hay shed in the middle of winter will find out how warm it can be. The wet hay starts to rot and gives off heat, which is kept inside the bale. Eventually, the hay gets so hot that the dry hay in the bale starts to smoulder and if not dealt with will burst into flames. Hay burns very well.

I can remember occasions during my childhood when I would go into the barn months after the hay had been stacked and find it full of smoke. I would then have to pull apart a stack to find the offending bales. This can be very dangerous because the sudden inrush of oxygen into the stack can cause flames to leap out. More than once I have found bales that, when opened, were too hot to touch in the middle yet the outside remained at ambient temperature.

Joel Woolf
Cullompton, Devon, UK

It was interesting to read earlier letters discussing how damp hay leads to internal heating in haystacks. As children growing up on a smallholding in west Wales, we had to build hay ricks using either small bales or loose hay. Hay was always carted and baled after the dew had evaporated from the fields, but occasionally rain, or the threat of rain, meant we had to build the ricks sooner than desirable. So

occasionally there were some damp bales, and occasionally one of these got into a rick.

If there were any worries about damp hay, the rick was tested every few days for any heating in its centre. This was done by pushing a pole between the bales or through the loose hay, and then someone would stick their arm in to take the temperature of the interior of the rick.

Only once did we have cause for concern, when the interior was very hot. All hands immediately pulled the rick apart and removed what appeared to be a smoking bale. It was also damp and had inadvertently been tightly packed in. We discarded the offending bale, allowed the others to cool, and rebuilt the rick more loosely.

I recall that another mishap involving fire and hay occurred in an outlying field by a railway line. The hay had been formed into a continuous row to make baling easier, when a spark from a steam train set it alight. Fortunately it was spotted, the burning hay was isolated and neighbouring householders brought water to damp it down, so only a small amount was lost.

Mary Sinclair
Narberth, Pembrokeshire, UK

Insurance companies in the 18th century insured farmers' haystacks in England, Scotland and Wales, so they clearly believed fire was possible. A standard phrase in the policies reads: 'Free from loss on such hay or corn as shall be destroyed or damaged by its natural heat.' This clause appears hundreds of times in the policies of Sun Fire Office in London.

Derek Morris
By email, no address supplied

The previous reply stated that insurance companies in the 18th century were willing to cover farmers' haystacks, implying that the insurers clearly believed hay could catch fire spontaneously. Actually, it doesn't imply spontaneous combustion of hay is possible or that insurers believed it to be.

Instead, the companies may have offered this insurance in the belief that it was not at all possible for hay to catch fire like this. They would thereby attain insurance-company nirvana: collecting a premium to cover a non-existent risk.

Ian Cargill
Leatherhead, Surrey, UK

Smelling double

Why do we have two nostrils?

Eleonoré de Bonneval
London, UK

Most animals, and many plants, are bilaterally symmetrical. The evolutionary origin of this trait can be traced back hundreds of millions of years. Our two nostrils are a consequence of this process.

What is much more interesting, though, is that there is little symmetry between the two nostrils, either in shape or function. For example, many people can smell different odours in each nostril, and you can test this yourself.

I have also noticed that when I have a virus, it usually only causes a blockage in one nostril at a time, which I would think has to do with the fact that killing the host is not usually in the best interests of an infectious agent.

Many organisms are not truly bilaterally symmetrical, however. Our heart and liver do not conform, for example, and some flatfish become asymmetrical in adulthood. But like everything else, there are exceptions to prove the rule.

Tony Holkham
Boncath, Pembrokeshire, UK

8 Health

Who dose?

If there are more than 200 different viruses capable of causing the common cold in humans, is it possible to catch several colds at once, with each one being caused by a different virus? Does immunity against one virus work against any of the others?

Leigh Sprague
By email, no address supplied

As far as I am aware there has never been any research into whether we can be infected with more than one common cold virus at the same time, but I don't see why it shouldn't be possible. Our bodies are often invaded by more than one virus, which is why we can have a cold and a cold sore.

But it's unlikely that the same cell could be infected by more than one cold virus, as cells produce a substance called interferon when under attack, which protects them from further invasion until that infection passes. If the different viruses were targeting different cells, however, there appears to be no reason why this couldn't happen.

Rhinoviruses, responsible for up to 80 per cent of cases, infect only a few cells, and the infection usually involves only a small portion of the cells of the respiratory epithelium, leaving many cells available for infection by other viruses. So it seems it would be possible to catch several colds at once, each one being caused by a different virus targeting different cells. You would probably not be aware of such simultaneous infections, though you may feel worse than if you had only one infection, as more of the respiratory apparatus would be infected.

The body has basically two types of immune reactions to a viral infection: non-specific and specific. Once infected, your body will first use the non-specific system, which fights off the intruder using a general mechanism that is deployed against all invaders. If this is not effective and the virus persists, the specific system is used.

The specific system recognises the virus and produces specific antibodies. Recognition is usually based on identifying the complex molecules of the virus (proteins, glycoproteins or complex polysaccharides), otherwise known as antigens. Once the antigens are detected, the body teaches cells called B and T lymphocytes to fight them. Some of the new B lymphocytes turn into 'memory' cells which can survive for several decades, so the next time the antigen is detected, the system can respond immediately.

To answer the second question, it is the specific system we are concerned with, and whether immunity would work against other viruses depends on how closely related the viruses are. The more closely related, the more likely the immunity would work.

Thanks to **Marc Van Ranst**
Head, Laboratory of Virology
Catholic University of Leuven, Belgium

Benign research

Why do benign tumours stop growing? Isn't it possible that understanding benign tumours and why they are not life-threatening could help us to treat malignant growths?

Charles Congdon
Oak Ridge, Tennessee, US

There is really no such thing as a 'benign' or 'malignant' tumour. All cancerous, or neoplastic, growths exist on a spectrum. What we tend to mean when we describe a tumour as benign is that it is unlikely to spread to other parts of the body or metastasise.

Benign tumours tend to have a capsule that keeps them in situ, whereas malignant growths penetrate the basement membrane that separates different tissue types before spreading through the blood or lymph systems.

Although benign tumours do not metastasise, they are actually far from harmless. Their continued growth can put pressure on organs, nerves or blood vessels, and can also cause significant tissue damage.

Endocrine tumours can produce vast quantities of hormones, which can be very harmful and may even induce other cancers in the individual. So a benign growth can cause significant health problems.

Benign growths can also eventually become malignant. Hepatic adenomas are primary neoplasms of liver cells called hepatocytes that are considered to be benign. But a mutation in the gene for the protein catenin is linked to a greater chance of their transformation into a malignant state.

The longer any benign disease is allowed to progress, of course, the greater the chance of the disease ultimately becoming malignant, which is why there is no true distinction.

Sub-classification of tumours based upon gene expression

is important for accurate prognosis and treatment options. Be reassured that the genetic pathways leading to neoplasms are areas of a great deal of research and are already leading to better treatment of neoplastic disease.

Chris Starling
Biomedical Scientist
Institute of Liver Studies
King's College Hospital
London, UK

Nosey question

Why does my nose run in cold weather?

Joe O'Brien
Ruislip, Middlesex, UK

This is caused by condensation and evaporation. In cold air there is not much water vapour but warm exhaled air is almost completely saturated with water vapour from its passage over the warm surfaces of the lungs and airways.

When the exhaled warm and moisturised air passes over the surface of the nasal mucosa that has been cooled by the cold air on its way into the lungs, it condenses, just as it does if you blow exhaled air towards the colder surface of a window pane or a mirror. As the nasal mucosa cannot take up all the moisture that condenses on it, the nose runs to get rid of the excess. The water running out of the nose is clear and clean condensed water, and is not a sign of an infection.

In the Arctic this constant running nose is a nuisance until you learn either to avoid breathing through the nose or to breathe in through the nose and out through the mouth. In fact, the original Arctic mitten has a piece of soft fur on the

back to wipe up the surplus water running out of the nose. After the water has frozen, the ice crystals can easily be shaken off.

The same condensation and evaporation problems arise when you are eating warm soup. Here the moist and warm air from the ingested soup will tend to condense on the nasal mucosa and create the same condition. By carefully adjusting your breathing you may overcome the problem by regulating the direction of the airflow over the nasal mucosa, or by simply only breathing through the mouth.

Leif Vanggaard
Hellerup, Denmark

Apart from being the sense organ involved in smell, the nose is the main route by which air enters and leaves the respiratory pathway. Before air reaches the lungs it needs to be warmed, moistened and cleaned and all these conditioning processes begin in the nose.

As those of us who suffer from frequent nosebleeds are only too aware, superficial blood vessels in the nose serve to warm the air. If you peer up someone's nose (perhaps do this when no one else is looking) you will see a number of large black hairs, which serve to filter out larger particles of dirt. Mucus secreted by glands lining the nose, aside from moistening the air, also traps smaller particles. The epithelial cells lining the nose have small hair-like structures called cilia on their surface, which beat to move the resulting mess towards the back of the throat where we swallow it, although the uncouth may choose to spit it out.

In cold weather the cilia on the epithelial cells beat less efficiently and the mucus dribbles out of the front of the nose rather than being shepherded backwards.

Similar ciliated cells line the windpipe, or trachea, where, along with mucus secreted by goblet cells, they trap dirt that has got through the nose and mouth. These cilia beat to move the material up towards the throat.

One of the reasons not to smoke is that it destroys these cilia, resulting in smokers developing a characteristic cough needed to move the impurities away from the lungs' respiratory surfaces.

Ron Douglas
Saffron Walden, Essex, UK

Lemon aid

Why does lemon juice make tiny cuts sting so much?

Patrick McGee
Glasgow, UK

It's the acidity. The skin of fingers is especially populated with pain receptors. These are particularly sensitive to pH, and lemon juice at pH 2.3, or even sometimes as low as 2.0, is sufficiently acidic not only to be detected but also to damage tissue by denaturing proteins. This has culinary uses in tenderising meat by hydrolysing tough collagen fibres, in preventing short-term browning of fruit by denaturing the enzymes responsible, and in cutting down the smell of fish by converting amines to salts.

The pH of a given acid depends on how concentrated the acid is, which at up to 6 per cent in lemon juice makes it the most acidic of the citrus fruits. The pH of blood and tissue fluid is around 7.4 – slightly alkaline.

The acidity league table of common foodstuffs runs something like this: lemon/lime juice pH 2.3, white vinegar 2.4, grapefruit 3.0, orange 3.5, yogurt 3.7 minimum, tangerine 3.9, tomato 4.5, and milk around neutral at 6.6.

However, lime and vinegar are less directly handled in western kitchens, unlike lemon juice, which can also be used as a household polish, cleaner and deodoriser, thereby increasing the likelihood of skin contact or surprise squirts in the eye, whose populous receptors sting like crazy to drive prompt prevention of damage. Pain receptors are equally sensitive to strong alkaline substances, notably soap at pH 10.

Gloves and goggles would help, but if you get juice in your cut, at least be relieved that its antibacterial action will lessen the risk of infection.

Len Winokur
Leeds, UK

Salt harsh

'Rubbing salt into the wound' was a way of preventing infection. But how did it work?

Peter Hallam
Derby, UK

Applying salt to a wound creates a highly saline environment, one in which it is difficult for microbes to grow. The high concentration gradient between the salt solution and the fluid inside bacterial cells makes it hard for the microbes to extract water from the solution without using a lot of energy. As a result, the bacteria become dehydrated and cannot function normally or proliferate.

Concentrated sugar solutions also have a dehydrating effect. This accounts for the extended shelf life of chutneys and preserves, and explains why honey can be used on wound dressings and, ironically, on bee stings as an antiseptic.

Scott Somerville
Cleghorn, South Lanarkshire, UK

Blood is 83 per cent water. Because salt is hygroscopic, it absorbs water, accelerating the tendency for blood to clot and drying the wound. This helps deny microorganisms a favourable habitat. Saline also generates osmotic pressure – it forces water out of microorganisms to equalise the salt concentration across their cell membranes. This can kill them, so salt acts as a disinfectant.

The stinging of the wound signals that salt does cause injury to the body. But in the absence of a better option at the time, killing a few healthy skin cells was acceptable collateral damage when the alternative may have been infection and possibly death.

Mike Follows
Willenhall, West Midlands, UK

Eyes right?

Why, when I'm tired, does my eyelid sometimes flutter?

Eileen Simmonds
Stockport, Cheshire, UK

The involuntary muscle contractions that ripple across an eyelid are called myokymia. They appear spontaneously and usually disappear again within a few days, though they can persist for up to three weeks. They are triggered by stress, fatigue or eyestrain, and can be exacerbated by caffeine or alcohol.

Readers familiar with the series of *Looney Tunes* cartoons featuring a coyote in pursuit of a roadrunner may recall that the earliest cartoons apparently showed Wile E. Coyote's facial muscles twitching to convey the stress of being outwitted yet again. The animators were taking poetic licence

because, though people with myokymia are acutely aware of the twitching and might even see the fluttering motion in their peripheral vision, it is virtually imperceptible to an observer, as can be confirmed by looking closely at the offending eyelid in the mirror.

Myokymia is quite different to the erratic blinking that may be a sign of anxiety, particularly in young people. This anxiety can also be manifested as shoulder shrugging and mouth twitching. Sometime between their 7th and 14th birthdays, about 1 in 3 boys and 1 in 10 girls will experience this anxiety and its symptoms.

Intense, longer-lasting and more widespread shaking of the body might indicate caffeine poisoning, alcohol withdrawal, an overactive thyroid gland or Parkinson's disease.

Mike Follows
Willenhall, West Midlands, UK

Age marker

Elderly people sometimes develop what are known as liver spots, which are darker areas or blemishes on their skin. Why, and what causes them?

Terry Anderson
Carlisle, Cumbria, UK

Tissues such as hair and skin do not produce their own colour; they get that from neighbouring cells called melanocytes through a remarkable process that involves injection of the pigment raw material, melanin, into the cells of the growing tissue.

Fresh melanin is practically colourless, gaining its final colour as the new tissue matures. Even after that, sunlight

may cause it to darken further, which is the temporary tanning effect we see in our skins after unaccustomed exposure to strong sunlight.

People, particularly those with a certain genetic make-up and who have exposed themselves to a lot of ultraviolet light as they aged, may find some of their melanocytes multiplying or darkening, or both, forming visible patches in the skin tissue. The resulting effect is known as freckles, lentigo, or naevus depending on the details.

Unlike most similar-looking medically important growths, the typical liver spot is flat, harmless and requires no treatment unless for cosmetic purposes. A medical check-up every year or two, in case a patch is hiding – or developing into – something more serious, is not a bad idea, but generally liver spots are of little medical interest except to dermatologists.

Jon Richfield
Somerset West, South Africa

We are still unaware of why they are called liver spots – Ed.

Finger fightback

I got a tiny cut on my finger last night and within half an hour it was warm and it hurt. The following day it was fine but it set me wondering how fast our bodies can detect an infective agent and slam into defensive action. Can anybody answer this?

Jana Svoboda
Corvallis, Oregon, US

Tissues, such as the epithelium found in skin, have immune cells within them called macrophages. These specialised cells can remain in the tissue for years and are activated by tissue damage or invasion by infectious agents. When this occurs, the cells secrete cytokines which induce inflammation, causing redness, swelling, pain, heat and loss of function in the tissue. This is to help destroy potential pathogens, form a margin around affected areas and recruit more immune cells. The immune response in this regard is instantaneous, although it is also fairly non-specific and the body takes longer (about three days) to start initiating a more specific response to a pathogen.

Jack Leitch
Portsmouth, Hampshire, UK

The finger may not have been infected at all. After injury, the body mounts an inflammatory response to initiate healing. The classic signs of inflammation are *rubor* (redness), *dolor* (pain) and *calor* (heat), as recognised by the ancients and experienced by your correspondent.

Andrew Cooper
Walls, Shetland, UK

Weight of expectation

Why, upon waking in the morning, am I usually rather desperate to defecate within about 10 minutes of rising?

Ally Portman
Manchester, UK

This is too individual a phenomenon for a reliable general explanation. It depends on personal physiology and health, your eating, drinking and sleeping habits, the regularity of other daily habits, any laxatives you might be taking, the nature of your food, your enteric worm load and a whole lot more besides.

If your diet is such that your stools are inclined to be on the loose side, they tend to settle in your colon at night, minding their own business, so to speak, quietly yielding their load of vitamins, minerals, amino acids, water and so on, to your system. However, by getting up you disturb their repose and their sloshing around disturbs your colon, stimulating it to motion. Enough said.

You do not say whether this is a nuisance, how long it has been going on, and whether it is associated with an unhealthy colon. If you are healthy and it is no worse than a scheduling problem, then just be grateful and make some appropriate allowances; millions of people would pay all they could afford for such convenient regularity.

Antony David
London, UK

Rate of escalation

In an attempt to lose weight, not only am I eating less, I am also exercising more by climbing a hill regularly. In terms of weight loss, is there any difference between climbing to the summit as fast as possible or at a more sedate pace with stops to catch my breath?

Alistair Scott
By email, no address supplied

Naively, you could say that a certain amount of work is done in moving from A to B, regardless of the rate at which this is done. I often see it quoted that x kilometres of walking or running will burn off the calories in one cream cake or whatever. This totally ignores the after-effects of exercise on your metabolism.

During and after vigorous exercise, the extra intensity of the effort raises body temperature which, in turn, raises the body's metabolic rate and makes it burn calories much more quickly.

Exercise also causes fatigue, by depleting the muscles' reserves of high-energy phosphorylated molecules and glycogen fuel, and by breaking down some of the protein structures that cause muscle fibres to contract: actin, myosin and the proteins that support them. During recovery these have to be replaced, and the body also builds extra to accommodate future efforts. This is called overcompensation.

Mitochondria in muscle fibres multiply to keep up with the extra demand for energy. Myoglobin, which transfers oxygen from haemoglobin in the blood to the mitochondria, also increases, and the number of capillaries carrying oxygenated blood likewise increases. A host of biochemicals, hormones and intermediaries have to be produced to enable all of this to happen.

This rebuilding takes energy, calories, materials – proteins from food that would otherwise be broken down for fuel – and the excess fat that you want to lose.

This overcompensation phenomenon is used by athletes in sessions called interval training, in which hard exercise is followed by lower-intensity work to allow recovery. A fairly long session of intense work sets off an enzyme called AMPase that turns on genes to improve the cardiovascular system. Very intense work, as in weight training, sets off another enzyme called mTORC that leads to muscle building.

If you wish to lose fat and improve your health generally, run up that hill, or such part of it as you can ascend at a stiff pace, not just once but a number of times. Repetition lets you keep up a higher intensity of effort for longer, which will have greater benefit. Walk down in between. Start with a couple of runs, then steadily increase that number. Give yourself a day

or two between sessions, but perhaps walk up the hill to have some exercise on your days off. You may feel hungry afterwards, so have a drink of low-fat milk, which will provide some immediate refuelling and will take away the hunger pangs.

However, be sure to warm up with a little jogging and stretching – no bouncing though, which can be a 'jerk too far' for tendons and ligaments. Finish with a bit more jogging and stretching as a warm-down.

Don't try too much too soon and be aware that if you start unaccustomed exercise, you may experience delayed-onset muscle soreness (DOMS). This is not serious, and will disappear after a few sessions. Keep moving, do light exercise and stretching. DOMS is like renovating a kitchen: you need to do a little internal demolition before rebuilding. You are feeling the effects of the initial demolition within your muscles.

Robert Bright
Amateur Rowing Association, silver level coach
Bedford, UK

Exercise is good for you, but it is a lousy way to lose weight because you have to do a great deal of it to burn up those calories. To burn off just one slice of buttered toast you would need to walk briskly for half an hour, or run for 15 minutes. That's equivalent to a 7.5-hour run for a kilogram off your weight. Regular exercise can raise the body's metabolic rate, so the average amount of energy you use will increase, prolonging the calorie-consuming effect, but it will not reduce your weight if you eat more to compensate.

Exercise also has lots of other beneficial effects that are well publicised, and vigorous exercise has even more, so do run up that hill. But remember that significant weight loss comes only from moderated food intake.

John Davies
Lancaster, UK

Pit problems

My doctor says that skin tags – small growths on the skin's surface – commonly form in the armpit. Why do they appear?

Ann Garmin
Glasgow, UK

Skin tags, the small projections of skin called cutaneous papillomas or acrochorda, are sometimes thought to be caused by friction because they are most common in areas of the body that receive most friction, such as armpits, the back of the knees, between the breasts and under them (especially where bras rub), and in the skin folds of obese people. The truth is that friction makes skin tags worse but doesn't cause them. People with type 2 diabetes must also put up with these annoying, unsightly but benign growths.

Most skin tags are small, but some can grow big enough to warrant removal. Doctors tend to 'freeze' them off with liquid nitrogen. Cautery and excision are also performed. Regrowth is common unless the root is completely removed, but it is worth noting that the scar can be larger than the original tag.

Skin tags are more common in women than in men and, like liver spots, are a side effect of maturity.

Toshi Knell
Nowra, New South Wales, Australia

Skin tags and warts are quite different; warts are caused by human papillomavirus and are infectious; skin tags are not infectious and appear to arise spontaneously. Tags are very common but are nothing to worry about. They are harmless in themselves and if they do cause irritation they can be removed by freezing or by tying off to stop the blood supply. It is not advisable to cut them off because larger ones can bleed profusely.

Chris Warman
Saltburn-by-the-Sea, North Yorkshire, UK

Heartache

Can heart muscles cramp in the same way as calf muscles? If not, why not?

Philip Welsby
Edinburgh, UK

In cramp, the muscle filaments known as actin-myosin units within the muscle are stuck in full or hyper contraction, in a state of energy depletion that may be exacerbated by electrolyte disturbance. The normal rhythmic contraction of the heart muscle – a phase called systole – pumps blood out of the left ventricle to the body. A systolic spasm or cramp would kill you very quickly. However, the heart is more likely to suffer serious disturbance to its rhythm rather than cramp, such as ventricular fibrillation, or asystole – when cardiac electrical activity stops.

But spasm of the smooth muscle of the coronary arteries can occur and cause a heart attack.

In Queensland we have the world's most venomous animal, *Chironex fleckeri*, a box jellyfish that lives along the coast and in the estuaries of the state. One of its toxic effects is to cause the heart to spasm in systole – it can kill you in minutes, and possibly in as little as 20 seconds. There is an antivenom but the chances of it working against a fatal dose are slim because you die before it can be injected. Immediate CPR should be attempted and I have seen it work in a marginal case.

Vinegar inhibits further release of the toxin, and prudent bathers in 'stinger' territory carry vinegar both to sprinkle on their fish and chips and, if necessary, to flush sticky tentacles from their skin.

Bob McCrossin
Cooroy, Queensland, Australia

Can the heart cramp like a calf muscle? Yes, if sustained muscle shortening is included in our definition of cramp. Such a contracture happens in severe metabolic inhibition, as in a heart attack. My group and I have shown that this contracture depletes cells of their energy currency, adenosine triphosphate (ATP).

We found this out by injecting a single heart-muscle cell with firefly luciferase, which releases light in the presence of ATP, while simultaneously watching the cell shorten – or cramp – in infrared laser light. (Thanks to the Biochemical Society, our paper is free at 1.usa.gov/Z4ZkX7).

During the contracture, myosin converts the ATP to adenosine diphosphate (ADP) which then spreads to the next cell. ADP initiates the cell shortening so the injury process propagates from cell to cell. We can speculate that preventing ADP from passing through the pores in the junctions between cells could prevent the spread of injury. To do this, I would suggest applying cytoplasmic adenylate kinase to the pores. This enzyme rapidly converts ADP to ATP. The question is how to do it. Like all science, one question begets more.

Peter Cobbold
Emeritus Professor of Human Anatomy and Cell Biology
Corwen, Denbighshire, UK

Pained response

I recently found a packet of ibuprofen that had expired. What happens to drugs such as these, and paracetamol, after expiry? Are they less efficacious or even harmful? They looked just fine although I thought it best not to take them.

Paul Instead
Grimsby, Lincolnshire, UK

We are grateful to Tony Holkham of Boncath, Pembrokeshire, UK, for pointing out that official medical advice is never to keep or take out-of-date medicines. Please bear that in mind while reading the following – Ed.

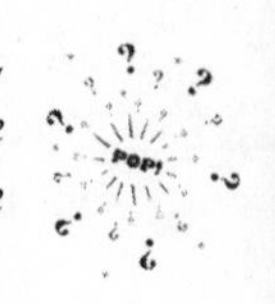

All medicines undergo stringent testing before they are marketed to ensure the active ingredient remains within an acceptable limit – usually at least 95 per cent of the amount stated on the label – throughout its shelf life, and that any degradation products are not more harmful than the original medicine. This shelf life is determined by carrying out stability trials on the medicine in its packaging.

Once a drug has expired, degradation continues and the content of active ingredient would be expected to decline. Tablets stored in bathrooms are likely to be exposed to more humid conditions and wider variations in temperature than when stored in a warehouse or pharmacy, so loss of the active ingredient may happen more quickly.

The appearance of the tablets is not a guarantee of safety, but any tablets that appear to be more friable than usual, broken, clumped together, 'growing' crystals, or otherwise not as expected should be considered suspect. In the case of aspirin, an odour of vinegar suggests the active ingredient, acetylsalicylic acid, has decomposed to salicylic acid and acetic acid. While salicylic acid is an effective painkiller, it carries a greater risk of internal bleeding than aspirin.

For out-of-date tablets, a risk/benefit analysis of whether or not to take them might include factors such as how far past the expiry date they are, how bad the pain is, and the availability of other pain relief. This does not constitute a recommendation, but an out-of-date pack of pain-relieving medication rejected during the evening may be very welcome at 2 am after several sleepless hours! Never take more than the stated dose – whether the pills are out of date or not.

Different considerations apply to liquid medicines. Not only are active ingredients more likely to degrade in a solution, the microbiological risks also become more significant. I am sure no readers sip cough medicine directly from the bottle in the bathroom in the middle of the night – but the consequences for future doses should be obvious.

Hillary Judd
Exeter, Devon, UK

Any drug must have an expiry date. However, this doesn't mean that immediately after this date the drug starts to undergo a degradation process and produce toxic by-products. Many compounds are highly stable if kept under good conditions, such as at room temperature, away from light and moisture.

A 2012 study in the *Archives of Internal Medicine*, in which various common prescription drugs were analysed between 28 and 40 years after their expiry date, revealed that drugs such as paracetamol (acetaminophen) and codeine contained more than 90 per cent of the initial amount of active ingredients, whereas compounds associated with aspirin were much more degraded.

Some drugs do form toxic products when they degrade. For example, tetracycline antibiotics are not very stable and can form degradation products that are toxic to the kidneys. So it all depends on the drug and the conditions of conservation. Some you can possibly take after the expiry date, but others you should avoid.

Olivier Sorg
Swiss Centre for Applied Human Toxicology
University of Geneva, Switzerland

Gum control

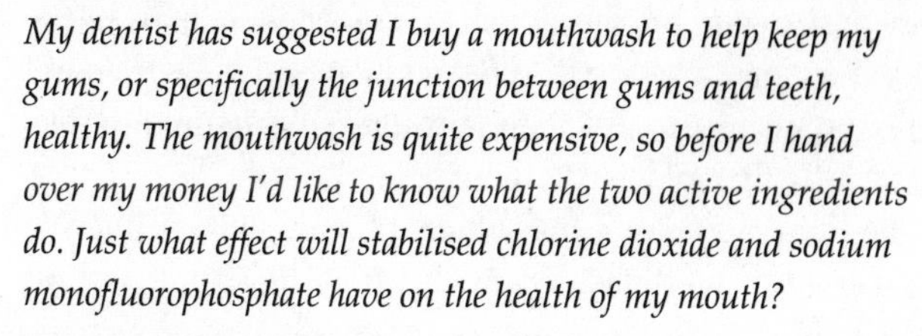

My dentist has suggested I buy a mouthwash to help keep my gums, or specifically the junction between gums and teeth, healthy. The mouthwash is quite expensive, so before I hand over my money I'd like to know what the two active ingredients do. Just what effect will stabilised chlorine dioxide and sodium monofluorophosphate have on the health of my mouth?

Janice Sturridge
Cardiff, UK

Chlorine dioxide is an antiseptic and sodium monofluorophosphate strengthens tooth enamel. But how do they perform in the lab? Raw chlorine dioxide gas decomposes violently in air, but is safe in mouthwash as a 5 per cent buffered aqueous solution with a pH of 9. In the US it is an approved and widely used disinfectant in food and water processing. It works by oxidation, which removes damaging chemicals but, unlike chlorine, it produces no harmful by-products.

Two other common antiseptic mouthwashes are hydrogen peroxide, which also oxidises, and chlorhexidine, which attacks a broad spectrum of microbes and can work through the soft plaque that clings to teeth and gums. But peroxide is a potential irritant that can aggravate symptoms. Chlorhexidine works by breaking down bacterial cell membranes but the brown remnants can discolour teeth, so it is unsuitable for frequent use. Chlorine dioxide also penetrates plaque, but being a more potent oxidiser it works at a lower concentration. And it kills a vast spectrum of bugs, including gram-positive and gram-negative bacteria, aerobes and anaerobes, viruses, fungi, spores, cysts and protozoa – all by dismantling their enzymes. Even bacterial spores with their multiple tough outer layers are dead in 5 minutes.

Tooth enamel is mostly made of hydroxyapatite, a complex of calcium, phosphate and hydroxide. Acids from food and sugars broken down by microbes strip away the hydroxide part, leaving the enamel more soluble, softer and hence prone to wear. Alkaline saliva provides hydroxide ions that gradually replenish it, though.

Sodium monofluorophosphate, like the sodium fluoride often used in toothpaste, provides fluoride ions that bind more tightly to the calcium-phosphate base of tooth enamel, making it more resistant to acid attack.

Len Winokur
Leeds, UK

Mouthwashes clean and disinfect the mouth and gums, particularly if you are prone to gum diseases such as gingivitis. Dentists recommend floss or interdental brushes, followed by brushing for 2 minutes. If this is followed by a mouthwash for 30 to 60 seconds, then any remaining bacteria will be significantly reduced, or eliminated.

Mouthwashes also have a significant cleaning effect, and will help to remove minor food traces that even determined brushing fails to dislodge.

Sodium monofluorophosphate protects tooth enamel from attack by bacteria that cause dental cavities. Stabilised chlorine dioxide mouth rinses contain a chemical called sodium chlorite, which is a salt used in the manufacture of chlorine dioxide. Both of these materials have an antibacterial role, but there is no point in buying expensive mouthwashes since cheaper products will also contain antibacterial chemicals and perform the same function.

Diluted hydrogen peroxide is a powerful cleaning and antibacterial mouth rinse, and is recommended by my own dentist. However, it tends to leave a metallic taste in the mouth. I get round this by adding about half a teaspoonful

of the concentrate to a standard quantity of the cheapest ordinary mouthwash. Having used this for about a year now, both myself and my dentist have seen a general improvement in my dental health.

J. A. Crofts
Nottingham, UK

I don't argue with dentists, but the ingredients mentioned are harmless, and the mouthwash should be dirt cheap. Monofluorophosphate supplies fluoride and inhibits some microbes, but many toothpastes contain plenty anyway; just don't rinse too soon after brushing. Chlorine dioxide (stabilised or not) is powerfully germicidal and compatible with bodily defences, but unless combined with silver it too is cheap.

Some 20 years ago mouthwash prices so insulted my intelligence that I checked their active ingredients and found that I had been paying way over the odds, so I bought some chlorhexidine concentrate and cheap vodka. Then I prepared a formulation of 0.2 per cent active ingredient in 6 per cent ethanol, adding a drop of clove oil and some saccharine tablets to mask the taste.

Since then a nightly rinse between flossing and brushing has worked miracles. I have no more ulcers, remarkable cleansing between teeth, no bad breath, and tartar reduction so drastic as to excite remark from successive dentists. Whitening toothpaste removes the harmless brown chlorhexidine stains. The cost is a small fraction of the price of mouthwashes that don't work as well as my own.

Jon Richfield
Somerset West, South Africa

My noise

Why does your own snoring usually not wake you up? Mine, I am told by those who hear it, can be especially loud.

Francis Curran
Johannesburg, South Africa

First, one snores most loudly when deeply asleep and hardest to arouse. We live in bodies so noisy they interfere with our reception of external information, and we are equipped to ignore our own noises such as breathing. We subconsciously subtract such noises from the signals we hear in order to deal with our world.

Our signal-filtering processes can have peculiar side effects, as anyone can tell when hearing a recording of their own voice. Not only is the sound unfamiliar, but even the accent. Sound cancellation enables us to sleep through our own snores, whereas a bedfellow's snore or the merest rustle might arouse us. However, even our own snoring awakens us if a recording is played back or an inadvertent grunt breaks its rhythm in a way our 'cancellation software' cannot neutralise.

Even without waking us, severe snoring commonly interferes with sleep quality because of noise and airway interference. Research shows that many snorers, not only adults, forfeit healthily deep sleep. Some decades ago, staff at the Red Cross War Memorial Children's Hospital in Cape Town, South Africa, showed that children suffering from snoring, or sleep apnoea, benefited from positive-pressure air supplies that held the respiratory passages open.

Antony David
London, UK

Back to front?

Sitting up straight using back and core muscles instead of slouching is better for posture and less likely to contribute to back problems. Why, then, is slouching more comfortable? How does it make sense biologically that the worse option is the easier?

Nick Brown
Oxford, UK

Tetrapods such as ourselves are segmented animals. This is an essential part of human morphology, the basis of much of our anatomy and our bodily function. One important consequence of this is that our spinal columns are flexible, articulated structures, rather than fused and rigid. This is most spectacularly demonstrated in snakes and fish, but is also notable in humans.

The spine combines compressive strength, leverage in various directions, flexibility and anchorage of muscles and tendons in order to create our broad range of movements. However, you often only appreciate this fully when you hurt your back.

To perform properly, the back requires fixed bracing by ligaments, as well as dynamic support and control from muscles; a back that could not slump also could not bend.

Beyond this, healthy animals, including humans, generally avoid unnecessary muscular exertion. A coursing cheetah in full flight bearing down on its prey is energetically extravagant, with every muscle in action, because losing its dinner would be a more serious matter than expending some energy. In contrast, a walking cheetah slouches efficiently.

And so it is with our posture; a healthy person in motion, whether walking or running, maintains good muscle tone, but relaxes their muscles when at rest and

relies on connective tissue to keep the skeleton from falling apart inside our skin.

Jon Richfield
Somerset West, South Africa

The difficulty of sitting upright compared with the 'worse' option of slouching is as much caused by poor furniture design as it is by anatomy.

When we stand, the muscles that flex and extend the hip are in balance with each other. This allows the body to assume its natural upright posture, with the pelvis tilted slightly forward and the lower (lumbar) spine curved inwards. Gravity acts through the hip and knee joints, so we can maintain this position with minimal effort.

When we move to a sitting position, the hips flex, causing the hamstring muscles to tighten and pulling the pelvis into a rearward tilt. This causes the lumbar curve to flatten and even reverse the direction of its curve.

The problem is exacerbated because the weight of the torso now passes behind the hips, tilting the pelvis back further. This slouched position feels comfortable, at least in the short term, because little effort is needed to maintain it – especially if you are supported by the back of a chair.

Sitting upright from this position requires a lot of effort to counteract gravity and the tightened hamstrings. Muscles in the lower trunk need to pull the pelvis upright, while muscles in the back haul the spine straight from its slouched shape. However, these muscles will soon tire and it will be more comfortable to go back to the slouching position.

Sitting upright is beneficial for breathing, digestion and general well-being. Some enlightened furniture designers now make forward sloping seats that facilitate this by reducing the tendency of the pelvis to rock backwards. And some seats do support the pelvis in its tilted back position while still allowing a fairly upright and functional posture.

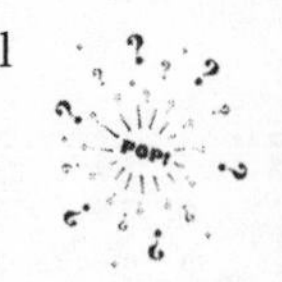

Chris Daniel
Llansanffraid Glan Conwy, Clwyd, UK

Zero G for me

I understand that the lines and sagging skin we acquire as we age are due to the sun and gravity. If I lived in a space station in zero gravity or microgravity away from the sun, would I stay looking young?

Mary Owen
Stoke-on-Trent, Staffordshire, UK

It is likelier that you would end up looking like a featureless bag of fat and fluid with porous bones and spindly attachments where your limbs and head used to be because human bodies would be free to expand in any direction without the pull of gravity.

Overexposure to the sun, especially at lower latitudes on Earth, can cause quite young people to look like leather saddlebags with eyes, but you don't have to go into space to avoid that look. Gravity on Earth, however, has little to do with making us look old; our bits sag once they begin to lose fibre, tone, or their internal padding after first stretching to accommodate fat, growth, or muscle.

In the hands of a competent surgeon, facelifts and injections of bacterial toxins can compensate for the stretching, and tissue implants can restore volume, but such operations can do nothing about the loss of tone or muscular function in general.

Gravity is just one thing that shapes our structure and

can't be blamed for the consequences of the other abuses our bodies suffer over our lifetimes, any more than you can blame food for obesity.

Antony David
London, UK

Living on a space station would be no picnic for your body. It's true you would be moving faster and so benefit from time dilation, according to Einstein. But a 6-month stint would only leave you 7 milliseconds in credit. Balance that against all the exercise you would need to do to fight muscle atrophy and the unavoidable loss of bone mass, both signs of ageing. If you were to venture through the Van Allen belts or orbit beyond them, you would receive an increased dose of ionising radiation from cosmic rays and the solar wind.

Mike Follows
Sutton Coldfield, West Midlands, UK

9 Cognition

Play it again, brain

Why do we sometimes get a tune or refrain stuck in our heads and play it over and over again even though it's driving us crazy?

Angelina Phipps
Bromley, Kent, UK

Getting a song stuck in your head is known by many different names including stuck song syndrome, or earworm, a term translated from the German word *Ohrwurm*.

Daniel Levitin, a neuroscientist at McGill University in Montreal, Canada, has suggested that it has an evolutionary origin. Before writing was invented, just 5000 years ago, songs helped people to remember and share information. Levitin suggests that variations of rhythm and melody provide the cues for easier recall, something that continues in communities with strong oral traditions.

This chimes with the findings of James Kellaris, professor of marketing at the University of Cincinnati, Ohio, who says earworms occur when you subconsciously detect something unusual about a piece of music. Usually between 15 and 30 seconds long, an earworm is replayed in your mind in a loop and is difficult to dislodge.

Music that is repetitive and simple, yet with an unexpected variation in rhythm, is most likely to become an earworm. For example, the repetitive melody and shifting time signatures of the song 'America' from *West Side Story*. Kellaris claims

that 98 per cent of people experience the feeling. About 74 per cent of earworms are songs with lyrics, 15 per cent are jingles from adverts, whereas instrumental pieces account for only 11 per cent of earworms.

Victoria Williamson, a music psychologist at Goldsmiths, University of London, suggests a number of triggers. Earworms are more likely to be bits of songs that have been heard recently or repeatedly. A song associated with a stressful or stimulating experience is also a good candidate.

In the early 1980s, Myron Warshauer tried to exploit this by patenting a 'musical floor reminder system' in multi-storey car parks in the US. The system helped people recall which floor they parked on by associating music and murals with each one.

Mike Follows
Willenhall, West Midlands, UK

Why lie?

Lying seems to be pretty much standard human practice: it seems virtually everyone does it to a greater or lesser extent. So why do we hate being caught?

Paul Loker
Guildford, Surrey, UK

Depending on the situation, being caught lying can make you seem untrustworthy, unreliable or dishonest. These are negative character attributes that we try to avoid. The reason is probably that untrustworthy individuals are likely to be ejected from social groups which, for early humans who lived in groups like other primates still do, amounted to a death sentence.

Therefore our dislike of being caught lying is a hangover from our past. It serves as a reminder to avoid lying (or at least being caught) for one's long-term safety. Further evidence for this idea is that being caught lying tends to trigger the sympathetic nervous system – part of the fight or flight response – which results in us feeling flustered and embarrassed. This suggests that the stress and danger of being caught might result in ejection from the social group.

Lewis O'Shaughnessy
Salisbury, Wiltshire, UK

It's because we are socialised from a young age to think it is wrong to lie, and this is backed up with the threat, implicit or otherwise, of punishment when discovered (either by human or superhuman agencies), ranging from public shame to hell.

The details of the socialisation will affect whether it is the telling of the lie (always discovered by the superhuman) or human discovery of the lie that matters most to you, and the punishment anticipated. But it is impossible to work out the details of this process without highly inappropriate experiments.

This socialisation is a lifelong process, so the balance between fear of superhuman and human discovery, as well as between fear of social shame and religious hell, are liable to change during someone's life.

At a more practical level, one presumably lies for a purpose, so discovery means the purpose of the lie cannot be achieved. The degree of regret depends on the importance of the objective.

Sarah Lambert
Department of History
Goldsmiths, University of London, UK

Free falling

I was travelling down from the 10th floor of my office building in the lift when I bent over to fasten my shoelace. I was suddenly dizzy and disoriented and had to kneel down to stop myself falling over. I never have any difficulties travelling in lifts, but presumably this was in some way caused by the fact that I was descending. I tried it again a week later with much the same effect. Why does bending over in a descending lift make me dizzy?

Richard Martens
Brussels, Belgium

It is not uncommon to feel dizzy in a lift. This is a symptom of motion sickness and occurs when the brain gets conflicting messages from the systems regulating balance and spatial orientation: vision, the vestibular system in the inner ears, and the proprioceptive system of receptors in the muscles and joints.

The vestibular system comprises the otolithic organs, which detect linear acceleration, and the semicircular canals, which detect rotational movements – pitch, roll and yaw.

When you stand in a descending lift, there is a conflict between vision and the otolithic organs. This is enough to disorient some people and cause motion sickness. If a person bends over, there is also conflict between vision and the semicircular canals. To the semicircular canals, the act of bending is exaggerated by the lift's descent, which is not detected visually. This is like walking towards the prow of a ship along an internal corridor as the ship pitches into a trough. This could have been your tipping point (in more ways than one) into mild motion sickness.

If you were to bend over in the descending lift and, at the same time, shake your head from side to side, it would induce aspects of roll and possibly yaw, depending on how far you were to bend. This would cause further disorientation and increase the likelihood of you landing on your head.

David Muir
Portobello High School
Edinburgh, UK

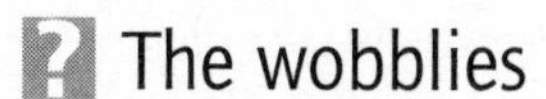

The wobblies

Why do my legs become wobbly when I stand near a cliff top?

Caroline Clarke
Dublin, Ireland

These effects vary from person to person. My wife, who is not acrophobic, gets a tingly feeling in her palms and soles, presumably in anticipation of a need to cling on with all the passion of her simian ancestors.

If however, like me, you have a poor sense of balance, and rely on visual input to supplement it, I speculate that this feeling is an effect of poor feedback. Normally one sways according to the balancing movements of the legs and unconsciously adjusts according to one's vision. Standing on a cliff where no level ground or rising walls offer visual clues to help you avoid toppling, your legs get poor feedback and keep over-correcting. That feels very insecure, which could set knees shaking.

When fear causes legs to tremble, the impulse seems a primitive one: small children and many other immature animals commonly throw themselves to the ground when frightened, in a state of helpless submission. Shaky legs might be part of that response.

A wry saying in Afrikaans expresses defiance of the symptom: 'Staan stil broek; jou baas is nie bang nie', which

translates into English as 'Stand still trousers; your master isn't afraid.'

Jon Richfield
Somerset West, South Africa

See the light

I was on holiday in Jersey and visited Mont Orgueil Castle, in Gorey. In one of the rooms there was a strange light display. The room was darkened and there were small spots of moving red light projected on the floor, walls and ceiling. There was also a vertical white light. It didn't seem particularly impressive until, as you glanced away from the vertical white light, in the corner of your eye it expanded briefly into a picture of Queen Elizabeth II. When you looked back, it was a simple vertical white light again. Not everybody present could see the picture, but probably around two-thirds did. What was happening, and how was the picture produced?

Pierre Castry
St Malo, France

The picture of the queen has been divided into vertical strips and these are projected in quick succession onto the same bit of wall. When you look directly at it, all you see is a bright vertical strip because each strip is overlaid on all the others. However, when you glance away, your eye is also moving so each strip hits your retina adjacent to the previous strip and your brain then fuses them into a single image of the queen. If you tilted your head through a right angle, you would have to flick your eyes up and down (relative to your head) to see the image and it would also be rotated through a right angle. The moving red lights wouldn't appear to have much to do with the queen's image, but perhaps are there to encourage you to move your eyes so that you see it.

Terry Collins
Harrogate, North Yorkshire, UK

The more interesting question is why some people saw the image and some did not. Your visual system has two different modes for moving your eyes around, each with its own control circuitry in the brain. The first is called 'smooth pursuit', a fairly self-explanatory name. The second is called a 'saccade', the motion by which your focus jumps from one place to another. You make these motions constantly – usually several times a second, as you move the high-resolution centre of your retina to point at different parts of a scene and build up a fully detailed picture of it. Normally you are unaware of these movements.

In fact, a great deal of neural processing power is dedicated to keeping you unaware of saccades so the brain can create a smooth picture of the world. Think of the lurching appearance of handheld video, and now imagine the camera is being deliberately jerked around to point at different targets, several times each second. One of the ways that your brain avoids the nauseating disorientation this would cause is to stop registering visual input during a saccade. This will usually prevent you seeing the image while looking directly at it. But peripheral vision is less strictly filtered because it's important to notice when something is approaching you, which is why you were able to see the image 'in the corner of your eye'.

The most likely explanation for why some people could see the image and some could not is that it's actually very difficult to make yourself look at something with your peripheral vision. When you direct your attention to something peripheral, your brain automatically plans and executes a saccade to it to get a clear, detailed view.

Steve Gisselbrecht
Boston, Massachusetts, US

The image is broken down into vertical stripes from left to right, which are shown on top of each other one at a time. You can see this if you can process information from the eye during a saccade. Many people's visual cortices can be fooled into seeing this 'mid-movement' information in dark surroundings.

I am especially interested in the phenomenon as I can do this easily in normal lighting but have yet to meet anyone else who can. It made the world an interesting place as a child: if a television programme was dull I could make the presenter fat by scanning my eye upwards or tall and thin by going in the other direction. It must be noted, though, that this only works with a cathode ray tube television. Fluorescent strip lights go from black to yellow to white and back again following the mains 50 hertz frequency when laid out like a stripy scarf in front of me and the LED boards in train stations break up into strange bundles of light that give clues about the display algorithm used to drive them.

Jeremy Richemont
London, UK

Taste revelation

The first time I tried an olive I disliked it intensely. The same thing happened when I first drank wine. Now I love both. What happens to our sense of taste to allow us to start liking such foodstuffs? And, just as intriguingly, what makes us persevere with them?

Tobias Montague
London, UK

As we age, we progressively lose parts of our senses of taste and smell. The first parts to go are those that make us dislike certain foods. Much later in life, we begin to lose the parts that allow us to enjoy our favourite foods. In between is the golden age of the gourmet – enjoy it while you can …

Alan Chattaway
Surrey, British Columbia, Canada

Our genetically acquired senses of taste and smell encourage certain preferences, such as that for sweetness, or avoidances, such as those for bitterness or faecal tastes, but we are by nature non-specialist, socialising omnivores. If we are neither to starve nor poison ourselves, we must learn which foods are good for us and which are not. A good rule is: 'What you or your friends don't know or don't like is nasty.' This rule is, however, subject to circumstances such as famine and social pressures such as hospitality.

The rule is also subject to experience, for example: 'That nasty-smelling cabbage left me feeling nice and full; now it doesn't smell so bad. But having vomited kidneys while in the middle of a fever, I can't even stand their smell!'

Alien foods tend to repel us at first, especially if our gastronomical experience has been narrow; a friend of mine offered an apple to an Ovambo girl in northern Namibia – she took one bite and promptly spat it out. She was used to wild fruit that, to western tastes, are insipid, astringent or generally vile. Perhaps some ancestor of ours was starving and so decided to try olives and then, having learned to prepare them and like them, passed on the liking. Now enthusiasts perennially offer us olives in various dishes, and gradually our system adapts.

Antony David
London, UK

Sharpen up

Why is the smell of pencil shavings so distinctive? I really like it, especially when they heat up as the sun comes through my classroom window.

Paul Keaton (aged 8)
Cardiff, UK

An important property of a good-quality pencil is that it doesn't splinter when sharpened. Red cedar wood has this property and was the wood of choice for pencil manufacturers during the 19th and early 20th centuries. Dwindling supplies of this tree forced manufacturers to turn to incense cedar in the 1940s and this is what many pencils are still made of today. Incense cedar has excellent 'sharpenability' and, as the name suggests, has a wonderful aroma.

Cedar trees have evolved to produce a cocktail of chemical compounds that includes cedrol and cedrene. This concoction can act as a protection against pests, bacteria and fungi. When a pencil is sharpened, this mix is released from the wood and the sun's heat increases the evaporation of the aromatics, enhancing the odour.

The limbic system of your brain is involved in aspects of olfaction, emotion and long-term memory. This helps to explain how, decades later, the evocative smell of pencil sharpenings can stimulate memories of your class and frame of mind when you first breathed in cedar's heady bouquet. A poor man's potpourri of pencil shavings and spices sitting on a radiator can elevate the spirit during winter's dark days.

David Muir
Science Department
Portobello High School
Edinburgh, UK

10 Alcohol

Ice spy

Ice in whisky produces lovely swirling patterns in the liquid. These are obviously to do with temperature and density differences between the melting ice and spirit. But what mechanism allows us to see these differences? What optical effects are at play?

Brian Higgs
Blackpool, Lancashire, UK

What your correspondent sees is caused by the mixing of liquids at different densities, but it can only be observed thanks to the refraction of light.

Light travels in a straight line through a vacuum and through a medium of uniform density. When light moves from one medium to another, its path changes direction by an amount that depends on the difference in density of the two media. The angle is calculated using Snell's law.

When light passes through a liquid of variable density, its path will change continually. This dynamic refraction results in an attractive shimmer effect.

To maximise this effect, allow your glass of spirit (or warm water for abstainers) to sit for a few minutes so the liquid is perfectly still. Hold the glass to a bright window and gently put in an ice cube.

You will see the ice melt and the cold, dense water flow around the ice in a laminar way. The flow then becomes turbulent as it moves down and mixes with the less dense

liquid below. The shimmering dynamic refraction allows you to see and appreciate the beauty of the flow patterns and the chaotic mixing.

An upside-down negative of this phenomenon can be seen if you hold the flame of a candle or cigarette lighter in front of a projector while it is shining white light onto a screen. The invisible convection currents created by hot, less dense air rising and mixing with the denser air of the room can be inferred from the shadows cast on the screen. Don't use a match because the mesmeric effect of the moving shadows may cause you to burn your fingers. In a still room, the shadows from the convection current can be up to a hundred times the size of the flame. Such dynamic refraction can also be seen in the heat haze above sun-soaked roads and above radiators by sunlit windows.

Dynamic refraction and its associated scintillation have even been immortalised in a nursery rhyme: 'Twinkle, Twinkle, Little Star'.

David Muir
Science Department
Portobello High School
Edinburgh, UK

Drug cocktail

Tonic water contains quinine and was originally drunk with gin in tropical climates in order to counteract malaria – but how exactly does quinine treat the disease?

Horace Laine
By email, no address supplied

Quinine is poisonous, but fortunately it is less harmful to you than to malaria parasites.

When feeding, the parasites break down haemoglobin in your erythrocytes, or red blood cells, releasing a toxic waste product called haem. Normally the parasites dispose of it by storing it in harmless insoluble clots. Although the precise way in which quinine and related drugs work is still obscure, it seems they interfere with the storage process, poisoning the parasites with their own wastes.

In appropriate doses, quinine does little harm to the human body, but it can have some nasty effects if abused. It can cause heart, nerve, eye and kidney damage, and pregnant women should certainly not take it.

The drug was once also prescribed for night-time cramps. However, tests of its efficacy have given such variable and sometimes alarming results that its use for this purpose is now widely discouraged.

Quinine is one of the most bitter substances known, and has become an ingredient in some bitters used in cocktails – but it is present in such small quantities that it probably amounts to less of a health hazard than the alcohol.

Jon Richfield
Somerset West, South Africa

Bottle it

Beer is usually packaged in brown bottles. Apparently this is because if it is packaged in clear glass, sunlight can damage its flavour. But how does sunlight damage beer? What chemistry occurs?

Roger Newton
Penzance, Cornwall, UK

I'm a student who has recently developed an obsession with home-brewed beer, so I hope I can offer some insight.

Nearly all beer contains hops. These provide the bitter taste and also act as a preservative (some natural deodorants even use hops for their antimicrobial action).

Hops contain isohumulones, which provide part of the bitter flavour – and this is where the problem with light arises. When UV light hits these compounds they decompose, leading to the creation of free radicals that react with sulphur-containing amino acids. The product is a thiol – the sulphur analogue of alcohols – and this leads to a 'skunky' flavour, so called for obvious reasons.

To protect against this, manufacturers use brown bottles that block out some of the UV rays. Clear and green bottles offer far less protection and so are more prone to develop a skunky flavour.

Try leaving a glass of beer out in the sun for even 10 minutes and compare it with one not in direct sunlight, and you will get an idea of what skunky beer tastes like.

Robert Law
Duffield, Derbyshire, UK

Visible and UV light causes a reaction in which riboflavin (vitamin B2) acts as a catalyst to break down bitter compounds called isohumulones that come from hops. This creates free radicals. One of these, 1,1-dimethylallyl, can then take a thiol group from a sulphur-containing amino acid to become 3-methylbut-2-ene-1-thiol. This compound is similar to two of the sulphurous compounds in a skunk's spray, namely 2-butene-1-thiol and 3-methyl-1-butanethiol. Brown bottles cut out most of the harmful wavelengths of light so beer does not become 'skunked'.

Eric Kvaalen
Les Essarts-le-Roi, France

Worm wouldn't?

Absinthe was supposed to have hallucinogenic effects due to the presence of extracts of the wormwood plant and has been banned in some countries at various times. Does this effect really exist?

Frank Arnhem
Amsterdam, The Netherlands

The hallucinogenic properties of absinthe are generally ascribed to thujone, extracted from wormwood. Thujone is bitter, with a slightly minty, aniseed-like taste, and is found in plants such as sage, oregano and juniper. However, because of absinthe's strength – sometimes more than 80 per cent alcohol by volume – and poor production techniques, the drink traditionally contained high levels of methanol. It is more likely that hallucinations, spasms and seizures attributed to thujone were caused by the methanol content, and that this led to bans on absinthe's sale, import and export.

Modern distilling techniques result in spirits with negligible methanol, if any. In the European Union, absinthe is usually between 45 and 75 per cent alcohol by volume and has a maximum legal thujone content of 35 milligrams per litre. As a rule of thumb, the higher the alcohol content and the darker the colour, the better the quality.

Dylan Brewis
London, UK

The history of absinthe is so fraught with urban legend and conflicting interests that no categorical answer to the question can be trusted. Even in the 19th century, the relative importance of the alcohol and the essential oils of wormwood (mainly *Artemisia absinthium*) in absinthe was much disputed.

Certainly thujone, the main active ingredient, is neurotoxic and can produce many of the symptoms of concern, and it is plausible that long-term heavy use might produce some of the alleged chronic effects. However, the likelihood of more modest drinking habits being harmful is bitterly disputed. Some people still blame van Gogh's mental decline on his consumption of absinthe, arguing that in those days the concentration of thujone was dangerously high, justifying its banning in some countries. Others assert that absinthe has hardly changed and is a benign intoxicant, stimulant, or both.

Antony David
London, UK

Absinthe is not hallucinogenic. I am drinking a glass of it as I write.

Absinthe contains the compound thujone which was once considered, on account of its molecular shape, to be part of a class of compounds called cannabinoids, which suppress neurotransmitters in the brain. But, in 1999, J. C. Meschler and A. C. Howlett showed this to be false (*Pharmacology Biochemistry and Behavior*, DOI: 10.1016/S0091–3057(98)00195–6).

Absinthe's reputation was generated by French winegrowers. Around the beginning of the 20th century, the French were drinking 36 million litres of absinthe annually and this was hitting the winegrowers' pockets.

The growers fought back with the aid of a burgeoning temperance movement, which bizarrely regarded wine as 'natural' because it came from the land. The winegrowers claimed that absinthe caused hallucinations, epileptic fits and suicides. They produced lurid posters and even created a medical-sounding term: 'absinthism'. A number of cut-price absinthe producers inadvertently aided them by using low-grade alcohol and a variety of genuinely toxic substances.

The anti-absinthe hysteria reached its climax when a vineyard worker in Switzerland, Jean Lanfray, shot his pregnant wife and two daughters before attempting, unsuccessfully, to kill himself. Public reaction focused on one detail: the two glasses of absinthe he had consumed.

The fact that Lanfray was an alcoholic was ignored. On the day of the attack he had not only drunk the two glasses of absinthe, but also downed a crème de menthe, six cognacs, seven glasses of wine with his lunch, another glass of wine before leaving work, a cup of coffee with brandy in it, an entire litre of wine after getting home and then another coffee with brandy. But such was the hysteria that people were in no doubt; absinthe was the cause. By 1908, the first ban became definitive.

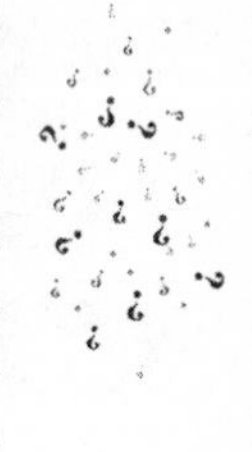

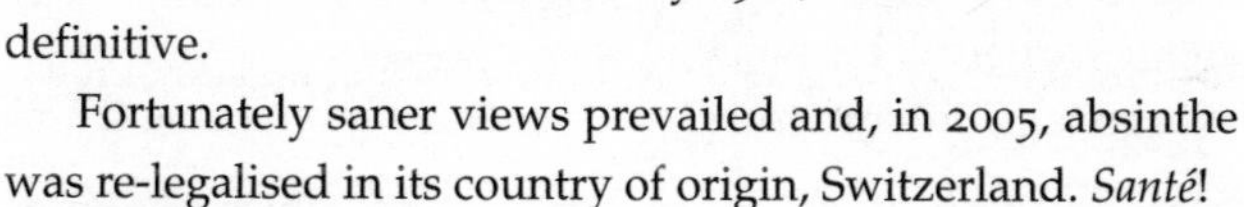

Fortunately saner views prevailed and, in 2005, absinthe was re-legalised in its country of origin, Switzerland. *Santé*!

Alistair Scott
Gland, Switzerland

P(ale) imitation

Why does alcohol-free beer taste so different from ordinary beer – and generally nowhere near as pleasant? Is it purely the presence or lack of alcohol? I'm not convinced that it is, because I added some neutral alcohol spirit to an alcohol-free lager we have here in the Netherlands to make it up to the strength of ordinary beer. My friends and I can report that it tasted absolutely foul.

So what else is missing from alcohol-free beer that makes it taste so different?

Anton Shaw
Delft, The Netherlands

Any fermented beverage is a complex mixture and includes a wide variety of molecules, such as microscopic particles of yeast cell wall or protein molecules.

Processes like pasteurisation change the taste so grossly that even untrained palates and noses can detect the differences. Other treatments might cause chemical changes or changes in the proportions of flavourful chemicals.

We can hardly generalise, because no two brewers use identical processes. However, producers of alcohol-free beer have two options.

First, they can simulate beer by mixing suitable components – but this doesn't usually yield good results. Second, they can produce a weak beer then extract the alcohol, replace any lost non-alcoholic components, such as organic acids, and re-carbonate the notional beer.

Having concentrated a good red wine by freezing it and then removing the ice, I can vouch that even this process, which is gentle in comparison with distillation, yields a thoroughly vile product. What is missing is more than a component, or the producer would simply be able to add it. What is lost is a physical and chemical state and balance.

Your 'neutral spirit', by the way, will still contain various incidental flavours.

Jon Richfield
Somerset West, South Africa

Bubbling under

I heard TV wine experts say that bubble size in champagne is a mark of quality. This suggests that wine-makers can control the size of the bubbles created when the bottle is opened and the carbon dioxide is released from suspension. Can this be true, or is the bubble size controlled more by the kind of glassware the wine is drunk from?

Polly Cawtherer
Oslo, Norway

The topic of bubbles in sparkling wines is a snare for the naive. Many variables affect the size and stability of the bubbles, and these variables conflict.

Excessive carbon dioxide concentrations generate large, frothy bubbles that suggest a cheap fizzy wine or too high a serving temperature. To give the bubbles the right spacing and lifetime – neither too frothy nor too dull – one needs the right content of surfactant molecules, releasing the bouquet without irritating the sophisticate's nose, while offering distracting tickles to juveniles. And the vintner can indeed influence the concentration of suitable molecules and microscopic particles in the wine.

Certainly the informed imbiber takes the glass into account, too. A champagne glass needs bubble nucleation sites, but not so many as to exhaust the supply of carbon dioxide, frothing over in a burst such as that caused by adding sugar. Traditionally, microscopic cracks in old glasses, or fibres from dishcloths, would supply starter bubbles. Now the bases of upmarket modern champagne glasses have minute laser-etched pits to nucleate bubbles in positions and sizes calculated to gratify people who are more concerned with the manner in which their fizz bubbles than why it does so.

Antony David
London, UK

Dissolved carbon dioxide can escape from champagne in a glass directly through the surface, which accounts for 80 per cent of the loss. This is why it's best to use a flute not a coupe. The remaining 20 per cent is lost through bubble formation.

Bubble growth occurs as excess carbon dioxide diffuses into the bubble as it heads towards the surface. The rate at which this happens, and thus the size of the bubble, depends on the amount of dissolved carbon dioxide in the champagne.

Champagne bubbles are produced by the 'méthode champenoise', a secondary fermentation in the wine created by adding sugar, yeast and yeast nutrients. This produces carbon dioxide. The key difference in sparkling products on the market is what happens after this secondary fermentation – the process of ageing. Non-vintage champagnes are required by law to undergo at least 15 months of ageing, while vintage wines require at least 3 years. Good champagnes undergo between 7 and 10 years and the greatest 12 years or more.

This ageing period means a loss of carbon dioxide through the imperfect seal between cork and bottle. This subsequently reduces bubble growth in the glass. Thus the better the wine, the smaller the bubbles – or perhaps more accurately, the more expensive the wine, the smaller the bubbles.

Peter Crawford
London, UK

Stirring stuff

What is the significance of James Bond's famous phrase 'shaken, not stirred'?

Is there really a difference in the taste of a shaken vodka martini, as opposed to a stirred one? And if there is, why?

Mark Langford
Stockport, Cheshire, UK

Just when we thought it was safe to go back into the cocktail bar, this old chestnut sprang back to life. In 2012, when we published our anthology Will We Ever Speak Dolphin?, *we thought we had it nailed. Shaken vodka martinis make the drink taste less oily. Many vodkas produced after the Second World War when Ian Fleming was writing his Bond novels were made from potatoes rather than grain (today's preference): because potatoes produce a distinctly oily vodka, James Bond liked his shaken rather than today's more acceptable stirred method. It appears, however, that there's more to it – Ed.*

It isn't only potatoes that lead to oily vodkas. After the Second World War, when grain was needed more to feed people than inebriate them, producers (legal and illegal) used any vegetable they could find. Root vegetables of many kinds – parsnips, carrots, beetroot and lots of others – make distinctly oily vodkas. And because they could still be used for distillation when they were going slightly off or even almost rotten, they were used in stills all over Europe. And all benefit from shaking when in a martini.

Vodka can be distilled from almost anything. A producer in England makes milk vodka. I can't vouch for its oiliness though, because I haven't tried it.

Alan Hamley
Salisbury, Wiltshire, UK

I heard that Bond's preference for shaking his martini rather than stirring it had nothing to do with chemistry. It was because shaking a cocktail involved only one of his hands, leaving the other free to grab his gun should any bad guys suddenly appear.

Joseph Kenny
London, UK

The truth is that Bond really should curb his intake. In research published in the *British Medical Journal* in December 2013, doctors in Derby and Nottingham in the UK calculated Bond's alcohol intake. Apparently he consumes on average the equivalent of five vodka martinis a day, or 92 units of alcohol a week. This makes him a 'problem drinker' – one

more likely to die from liver failure than lead a glamorous lifestyle, let alone hold a Walther PPK steady or manage as much sex as Ian Fleming's books suggest he does. In the novel *From Russia with Love*, Bond actually downs 50 units in one day. The researchers suggested he only preferred shaken martinis because he was already shaking as a result of his excessive boozing. They concluded that Bond should be 'referred for further assessment of his alcohol intake'.

Still, while it's an important point they are trying to make, the researchers surely understand that Bond is a fictional character. His level of alcohol intake is as unlikely as his ability to break into Fort Knox, fly a space shuttle or survive the more than 4500 bullets fired in his direction.

Charles Black
Basseterre, St Kitts and Nevis

Although medical researchers have concluded that Bond drank too much, they are clearly forgetting that he was a spy and used clandestine methods. Surely everybody knows that secret service operatives in the field only pretend to consume vast quantities of alcohol.

They fake becoming drunk so that they are well prepared for when somebody starts to blab their megalomaniacal plans, or the guns come out. Check the potted plants and umbrella stands near where Bond was sitting and you'll probably find that they contain olives and lemon peel, and smell of vodka martini.

Karen Jarrold
Milton Keynes, Buckinghamshire, UK

Apparently the third man to play Bond in the movies, Roger Moore, never uttered the words 'shaken not stirred' because he didn't think he could do justice to the way the first Bond, Sean Connery, delivered them. Ironically, before his Bond days, while playing the Saint in the eponymous TV series, Moore did use the phrase.

Colm Sheehy
Liverpool, UK

11 Eating

Straight and narrow

Why do some sausages curl when they cook – even those that start off straight? And why do some stay straight, such as hot dog sausages? And how can I stop the ones that do curl from doing so, allowing me to brown them evenly?

John Virgoe
Canberra, ACT, Australia

Sausage skin is mostly collagen, which shrinks violently in cooking because it is degraded and dried by the heat. Most cooking equipment applies heat asymmetrically, causing uneven shrinkage, which creates curl. Partly to avoid this problem, modern hot dogs do not have a collagen skin.

Brute force, by using skewers or confining the sausage in a barbecue grill clamp, will counter the problem. Subtler strategies such as severing fibres can forestall the effect in sausages and in steaks and chops. There are also tricks based on heating sausages evenly and slowly, or by judiciously turning them so that the convex side is always towards the heat. Such tricks ensure that the skin on opposite sides shrinks in unison and opposing pulls counteract each other.

Boerewors stands with rotary sausage grillers take that principle to its logical conclusion. Boerewors sausages can curl with the best, but the griller comprises a parallel array of closely spaced hot metal cylinders that roll continuously in the same direction. The sausages rest on two cylinders at a time and roll in the opposite direction. The products are sausages to dream of: big, tasty and straight, almost independent of the operator's skill. Eat your heart out bangers!

Jon Richfield
Somerset West, South Africa

And as our next correspondent proves, innovation is not dead – Ed.

We solve the sausage curl problem by inserting a pre-soaked bamboo skewer into the sausage to about two-thirds of its length, leaving about 10 centimetres of skewer for a handle. Then we barbecue them and eat them like a sausage lollipop which you can dip into sauce. They stay completely straight and are a hit at parties. You also have one hand free for your beer or wine.

Ian Jones
Carindale, Queensland, Australia

Turning to mush?

When I start to fry mushrooms in oil they quickly absorb all the liquid, making the pan quite dry. But after a couple of minutes they suddenly start to release it all again. What's going on?

Francis Keynes
Dover, Kent, UK

In fields, mushrooms seemingly appear overnight, generally following damp conditions. This is because the mushroom, or fruiting body, draws in water rapidly from its mycelium within the soil to expand and appear above the surface as a mushroom. As a consequence mushrooms contain about 90 per cent water and have very few calories.

What's more, unlike plants, which have cell walls containing cellulose, the cell walls of fungi contain chitin. When mushrooms are fried in fat, the fat is initially absorbed into the chitin wall which, because of the heat, then ruptures to release the water that has been contained within the cells.

How one overcomes this problem of mush is a matter of debate. I have drained off the fluid and added double cream to the pan. A friend who teaches domestic science told me that one should immerse the mushrooms in a hot stock for a few seconds and then fry them. This way they do not go mushy. I tried this and she is correct. I assume that the hot stock causes the chitin wall to become more flexible so that it does not crack and release the internal fluid within at the time it is being fried.

Finally, if you are lucky enough to come across a giant puffball (*Calvatia gigantea*) while foraging, a slice fried in butter out-tastes any mushroom and never becomes mushy.

Gillian Coates
Anglesey, UK

As ever with fungi, we advise readers to be absolutely sure of the species they have brought home before consuming it – Ed.

Dairy division

Sometimes I freeze milk. If I defrost it within two or three weeks of freezing it is fine. But if I defrost it after two or three months, it separates into a watery part and a thicker, white part. Why? How does it change during the time that it is frozen and apparently inert?

Peter Jones
Uxbridge, Middlesex, UK

Milk is a solution containing a suspension of particles – a structure that concentrates surprising quantities of insoluble materials into a liquid that normally remains stable until it reaches the stomach, where enzymes curdle it so that it stays put long enough for proper digestion.

Freezing creates a mesh of ice crystals, trapping particles of butterfat, membranes from cells, micelles (protein particles that hold astonishing concentrations of nutrients in suspension, especially calcium salts) and regions of imperfectly crystalline ice containing dissolved materials. This 'apparently inert' mass is anything but. Constant recrystallisation ruptures cells and micelles, driving dissolved materials out of the solution, and stranding proteins between the growing crystals, where they become distorted and denatured.

The materials trapped between the ice crystals become concentrated into less and less solvent (water). The process is slow, but fats, proteins and insoluble calcium phosphates from the ruptured micelles agglomerate, react chemically and entangle physically, congealing and resisting resolution or re-dispersal.

It is still nutritious and might be good in cheese or ice cream, but is too chewy to be pleasant in tea.

Antony David
London, UK

Juiced up

I work in a restaurant that has a bar with a zinc surface. We use lemon juice to clean the bar top, which leads me to two questions: why is lemon juice so good at cleaning zinc? And why does lemon juice on zinc smell so awful? We have to clean the bar in the early morning to avoid customers being put off.

Sean Key
London, UK

Is the bar counter quite old? Zinc used to be extracted from its ore, an impure zinc sulphide known as sphalerite or zinc blende, by heating it with carbon. The metal arising from this 19th-century process, last used in the 1950s, was contaminated with unconverted zinc sulphide and traces of heavy elements close to zinc in the periodic table, including arsenic.

When pure zinc is treated with acid such as citric acid from lemons, hydrogen gas is generated, and the oxidised or stained surface layer is dissolved, leaving a bright, clean finish. But the zinc sulphide in old zinc sheet will give off the toxic gas hydrogen sulphide, notorious for its rotten-egg odour. In addition to this, any arsenic will be converted to another poisonous gas, arsine, which smells of garlic.

Although the concentrations of these gases will be low, they are best avoided, so give metal polish a try.

The zinc-acid reaction was used to detect traces of arsenic in one of the first forensic tests, developed by James Marsh in 1836. This involved heating up material, such as exhumed body parts, with zinc and hydrochloric acid. The resulting hydrogen was burned in a gas jet aimed at a cooled metal or a glass surface, where it would leave a shiny black deposit if arsenic was present. But care had to be taken because impure zinc would give false positives.

The Marsh test was cranky and time consuming, and the 1922 case against English solicitor Herbert Armstrong was delayed because of disputes over who should foot the bill for the lab work. Armstrong had bought arsenic trioxide to kill dandelions in his lawn, but found it equally effective on his wife.

John Rowland
Derby, UK

Zinc is reactive, and the dull and dirty film you wipe off is a layer of zinc carbonate and oxide complex that forms on contact with carbon dioxide in the air. But this film does protect the underlying metal from further reactions. However, zinc carbonate reacts with acids to form a soluble salt and odourless hydrogen.

Lemon juice is particularly effective because of the low pH (or high acidity) of its citric acid. In a bar, it is the cleaner of choice because there is always a handy supply and because of its gastronomic safety and fragrance.

The bar will be subject to food and drink spills, residues of which lodge in minute surface scratches and pits. Microbes metabolise this bounty anaerobically to produce hydrogen sulphide – the characteristic smell of rotten eggs. This is mopped up by the oxide in the surface film in a reaction that converts it to zinc sulphide. But on cleaning with lemon juice, the sulphide reacts with the citric acid to yield hydrogen sulphide again: it is this that makes the zinc stink. Scrupulous hygiene doesn't really help because even 0.47 parts per billion will turn the nose up.

Len Winokur
Leeds, UK

Condiment cleaner

So lemon juice can clean a zinc bar top, but it's a smelly process. I too work in a restaurant and we have a copper bar top. We clean it with tomato ketchup, which works really well and, more importantly, doesn't produce a bad odour. So why does ketchup clean a copper surface without smelling terrible, unlike lemon juice on zinc?

Marcim Jakubowski
London, UK

The zinc in a metal sheet will not be very pure, and will contain sulphur impurities such as zinc sulphide. Because zinc is a very reactive metal, it rapidly forms an oxide layer on its surface, which protects it from further attack from oxidising agents, such as those in the air. However, under the moist conditions of a bar top, particularly under slightly acid or alkaline conditions caused by dropped food or drink splashes, zinc will tarnish quite rapidly. Cleaning with an acidic cleaner will restore the surface, but the zinc sulphide impurities in the metal will react to produce hydrogen sulphide (the smell of bad eggs). A final clean with warm water, followed by a dry polishing, is highly recommended.

A copper bar top is different. A copper sheet will tarnish on exposure to the air – which contains tiny amounts of hydrogen sulphide – by producing a layer of copper sulphide, which is black. This can be removed with any acidic cleaner, but little of the sulphur ends up as hydrogen sulphide, so there is no bad smell. Ripe tomatoes and ketchup contain malic acid, and have a pH range of 3.5 to 4.5, making them ideal (if messy) cleaning agents.

John Crofts
Nottingham, UK

Best before?

What allows fruit cake to survive so long? I had a piece of a friend's 2-year-old wedding cake and it tasted fine. If it had been a sponge cake, it would have mouldered away.

Penny Allamby
London, UK

It's the brandy!

No, actually it's the sugar content. Sugars are hygroscopic, that is, they draw water from their surroundings, including from any bacteria or fungi. This prevents growth of such microbial contaminants. Bees process honey for exactly the same purpose, reducing the water content of the collected nectar to around 20 per cent so that the honey, which is 80 per cent sugar, can resist microbial spoilage.

Jam – a method of preserving fruit in sugar – is about 60 per cent sugar. Traditional fruit cake can be around 60 per cent carbohydrate, with the dried fruit and sugar together contributing about three-quarters of that. In other words, fruit cake can be as much as 50 per cent sugar.

In the days before refrigeration this was an important means of preparing food for storage or travel, as well as providing a sweet delicacy. In Europe prior to the introduction of sugar from sugar cane in around 1100, dried fruits and honey were the only such preservatives, hence the ubiquity of fruit cake, fruit mince and the like in post-medieval European cookery for special occasions.

E. A. Smith
Canberra, ACT, Australia

Fruit cake can survive a long time because of its high sugar content, mostly derived from its dried fruit, which acts as a preservative. Fruit cakes are also cooked for a long time, ensuring high temperatures through the cake which means it starts out pretty much free of bacteria.

It is also dense, with little air space, which prevents it from drying out. The practice of pouring high-alcohol liquids over the cake at regular intervals acts as a steriliser and also ensures it remains a poor medium for microorganisms.

However, it is quite easy to make what is called a non-keeping fruit cake. I was once asked to make a wedding cake

to a particular recipe. I was quite concerned at the ingredients, which included finely cut raw pineapple, but I was assured that it was an old family recipe and worked wonderfully.

Duly baked, aged, iced and delivered, I heard that the cake was delicious – especially the green layer at the base. I did not hear that anybody suffered any ill effects but I can guarantee that the cake did not start out with green ingredients!

A sponge cake never reaches high internal temperatures for very long, if at all, and has a relatively low sugar content as well as little structural integrity due to being largely air. It probably won't go mouldy unless it gets wet, but it probably won't taste so good for very long.

Jan Horton
West Launceston, Tasmania, Australia

Chocolate medals

Apparently it's a curious fact that countries where chocolate consumption is high tend to produce more Nobel prizewinners. The correlation is quite stark. Are there any reasons why this could be the case, including statistical anomalies not directly related to chocolate? Ironically, my country, Sweden, home of the Nobels, was one of the very few that didn't fit the pattern.

Anders Peters
Uppsala, Sweden

Rich Western countries tend to have a history of academic success, as shown by the number of Nobel prizewinners they produce. They also have a high consumption of foods that are considered to be less healthy, including chocolate. Sweden, as your correspondent states, does not exhibit this association.

By examining international obesity league tables, we can see that Sweden ranks fairly low for a Western nation, suggesting that it has taken on board health messages regarding diet. Other rich countries with an academic pedigree, such as the UK, the US, Canada and Australia, have not.

In the 1980s, Sweden reacted to high cardiovascular disease mortality rates by introducing an official food-labelling system and a programme of health screening and counselling. This programme was reported by participants to be influential in supporting lifestyle change. The food labelling brought about a significant increase in the sales of low-fat products. It is a reasonable hypothesis that the actions of the Swedish government have caused the non-correlation between chocolate consumption and the number of Swedish Nobel prizewinners. The message should be obvious to us all.

An alternative hypothesis could be that non-Swedish academics consume massive amounts of chocolate.

David Muir
Science Department
Portobello High School
Edinburgh, UK

Correlation need not imply causation. However, correlated independent effects commonly have related causes, and the failure of some countries to fit the pattern suggests that this correlation is an example. Although I'm not a true chocoholic, I would love there to be a causative connection, but my own suspicion is that many of the achievements awarded Nobel prizes were nurtured in environments that rarely occur in nations lacking luxuries such as good nutrition, good health, good education, hubris, enterprise and being able to afford lots of good chocolate. As Lady Randolph Churchill said: 'We owe something to extravagance, for thrift and adventure seldom go hand in hand.'

Jon Richfield
Somerset West, South Africa

It cannot be true that the amount of chocolate consumed increases your chances of winning, because women have taken just 5 per cent of the Nobel prizes awarded. One need not accept the sweeping generalisation from the media machine, which would have us believe that women love chocolate more than men do, but it is probably fair to say that women and men enjoy chocolate in roughly equal numbers.

Richard Fu
London, UK

Knobbly cheese

Why do Grana Padano, cheddar and feta cheeses, to use three random examples, all break in different ways and have inside surfaces with very different appearances? After all, they are all cheese.

Maria Brigida
Pesaro, Italy

Don't forget Brie, which starts out as a delicious cream cheese and ends up as a delicious, creamy cheese. Cheese is mainly protein and fat, with combinations of water, salts and organic products but, as with every food, the recipe and proportions of ingredients make the difference between a feast for a king and indigestion for a beggar. Also, the character of a cheese changes as it matures.

Fats, proteins and water determine cheese texture. The protein molecules start out as chains and, if they pack neatly with little water and fat between them, they make a strong plastic – hitting someone with a block of Grana Padano could kill them. In contrast, bacterial and fungal enzymes break up proteins in soft cheeses like Brie until they cannot form strong structures, and fresh cottage cheeses have not yet had a chance to form such structures at all.

Soft, matured cheeses also contain a great deal of fat, which forms a creamy emulsion with the water and proteins. There are enough such combinations of ingredients and recipes to make thousands of cheese types.

Antony David
London, UK

In the 1970s and 1980s, I was at the UK's National Institute for Research in Dairying, and I did a lot of research into the science of cheesemaking.

Grana Padano is a hard cheese, cheddar is semi-hard and feta is a soft cheese. Their differences in texture and flavour reflect differences in their production, which change fat and moisture content and determine how much curd structure is left in the final product. Grana Padano starts with partly skimmed cow's milk, cheddar uses full-fat cow's milk and feta uses full-fat sheep or goat's milk.

In cheddar cheese, it is possible to see the boundaries between curd pieces where they have been cut up in the cheese vat. If you break a piece of mature cheddar gently, it will break along these edges. Moisture in the cheese acts as a plasticiser, making it pliable and elastic.

Fat plays a vital role in softening the cheese, acting as a lubricant between the grains as well as giving a smooth feel in the mouth. The making of Grana Padano causes more fusion of its curd pieces, giving a closer texture. The lower moisture and fat contents make it harder than cheddar.

Nearly all cheesemaking starts the same way: a coagulating enzyme is added to the milk, along with an acidulant.

Usually, the acid is generated by lactic acid bacteria and the enzyme is rennet. This is often extracted from calves' stomachs, though microbial rennets are also available. At the

near-neutral pH of milk (about 6.7), rennet acts on casein, the main group of proteins in milk, causing it to coagulate. As the bacteria multiply, the acidity increases and the curds shrink.

To release the whey, the curds are cut into pieces – about 2 cubic centimetres for cheddar and around 0.5 cubic centimetres for Grana Padano. For feta cheese, the curds are cut up and placed in a mould, where they are allowed to drain.

In cheddar cheesemaking, the curds and whey are warmed to 36 °C for an hour or so. This warming – called scalding – shrinks the curds and squeezes out more whey. The bacteria continue to produce acid and the pH falls to around 5.5. At this point, the curds form a semi-solid mass that is cut into blocks.

Next, these are compacted by stacking, or 'cheddaring'. The weight of the blocks on top compresses those underneath, and the curds fuse together. By regularly turning the pile, they are all treated equally.

This continues until sufficient acid has been produced and the blocks have a structure that resembles chicken breast meat. They are milled into much smaller pieces and salt is added.

Finally, the curds are placed into moulds and pressed. The cheeses then ripen over several months, which is when the flavour develops.

In Grana Padano cheesemaking, the curds and whey are heated to 55 °C. The curds are then placed in a cheesecloth to drain, cut into two pieces and put in separate moulds, where they stay for about 8 hours. Then, in new moulds, they spend the next three weeks in brine. The maturing process takes another nine months at least.

In feta making, the curds drain for several hours. They are cut up, salted and left for several days while the salt infuses. They are then placed in brine for several weeks at 20 °C, before being transferred to a cold store for final maturation. The result is a smooth-grained, high-moisture, salty cheese.

Richard Marshall
Senior Lecturer
Food Enterprise
School of Society, Enterprise & Environment
Bath Spa University, UK

Double your luck

My supermarket sells eggs that are guaranteed to have double yolks, but this is surely a random process. What are the chances of a double yolk, and how do egg producers ensure there are enough to fill their boxes? Can they engineer them via artificial means? My local supermarket always seems to have a plentiful supply so presumably it's easy enough to find or generate them.

Patrick Ness
Maidstone, Kent, UK

The process is not as random as it first seems; the most common cause of double-yolked eggs is when two ovules are produced so close together that the hen's oviduct processes both yolks into the same shell. The tendency to produce more than one ovule at a time is influenced by genetics, so some breeds of chicken produce far more double yolks than others.

Certain cross-breeds ovulate rapidly, so that most eggs are double-yolked (and incidentally only rarely produce chicks). Some people prefer double-yolked eggs, regarding them as lucky, rather like four-leaf clovers. This is also reasonable from a nutritional point of view because most of the value of the egg is in the yolk, despite its high cholesterol level.

However, in some regions people prefer single yolks. This

offers an incentive to market single or double-yolked eggs at a premium on the basis of regional tastes. This is aided by candling – you can shine a light through the eggshells to tell whether they are internally unusual, containing blood spots, double yolks or the like, and then sell them according to consumer preferences.

Jon Richfield
Somerset West, South Africa

When young hens begin to lay eggs, their ovulation is irregular and it is not uncommon for them to shed two yolks so close together that these get wrapped up in the same shell, giving a double-yolked egg. These are obviously larger than normal eggs and will be filtered out when the eggs are graded. Inspecting them over a bright light will confirm the double yolk.

When the birds settle into a regular laying cycle, such eggs become rare. Your questioner's supermarket presumably purchases its eggs from a large producer that always has a proportion of young birds.

J. Allen
Grantham, Lincolnshire, UK

? Seeing red

Why do some shellfish turn red when cooked?

Tina-Louise Brown
Algiers, Algeria

Certain shellfish, such as lobsters, turn red when cooked because they are red to begin with – we just can't see it. In life, the red pigmentation in their shell, created by the presence of astaxanthin, is combined with a range of proteins and other pigments to produce the dull camouflage colours that enable the crustaceans to blend in with their environment.

When you boil a lobster, or a prawn or shrimp, proteins in the shell denature and unwind, releasing their attached pigments, but while the others break down at high temperatures, astaxanthin retains its stability and reflects light at the red end of the spectrum. This may seem ironic, given that astaxanthin belongs to the class of carotenoids known as the xanthophylls, which literally means 'yellow leaves'. Astaxanthin provides not only the colour for cooked crustaceans but also for red salmon.

Astaxanthin is used as a food additive for farmed salmon to make their flesh resemble the line-caught variety, and is also fed to intensively reared hens to give their egg yolks a richer orange colour. Although it has been well established that astaxanthin poses no health risks, this practice has aroused controversy because it is felt that consumers might be misled into thinking they are eating an organic, hence arguably 'healthier', product.

Although there is no evidence that battery hens' feathers turn pink after eating astaxanthin, the pigment can make its way through the food chain into certain birds' plumage. The most famous example of this is the flamingo, which can obtain its rosy hue from astaxanthin in the shells of the shrimps that it sieves from brine pools.

The corollary to a boiled lobster being red is that an unboiled lobster is not. Lack of redness usually means that it's still alive, even if it has been out of the water for some time. When I was young, my parents took me on holiday to Brixham in Devon, UK, where my father would go fishing. One day we met him brandishing a huge mottled blue-brown lobster. Spurred by some inner instinct, I vehemently maintained it was still alive. The adults assured me it was not.

We returned to our cottage where the lobster was placed on a worktop, and I was left to watch television in the darkened living room while my parents went to buy salad. Moments after they left, I heard a distinct 'click-clack' noise coming from the direction of the kitchen. Then I heard a heavy thud, clicking and shuffling. Silhouetted in the kitchen door was a large lump on the floor, brandishing two massive claws in the air above it.

Being of that generation who learned by watching *Doctor Who* that the best way to deal with incipient peril is to seek safety behind the nearest armchair, that's what I did. Unfortunately, this was beside the TV, and following millions of years of instinct, the monster from the deep began shuffling towards the only source of light.

Just as it made it to the chair behind which I was cowering, I heard the key turn. My parents were greeted with a screech of 'I told you it was alive!' My mother grasped the crustacean by the tail and whisked it off to a saucepan. Ever since, I have never trusted a lobster that wasn't as scarlet as a guardsman's tunic.

Hadrian Jeffs
Norwich, Norfolk, UK

12 Transport

Burning down the road

During the past 100 years, humans have been burning oil, natural gas, peat and coal. In the next 50 years we will burn even more. Burning hydrocarbons produces carbon dioxide and water. How much has this water added to sea-level rise?

Alfred Jacobsen
The Netherlands

The cumulative amounts of oil and gas that have been used globally are not known with great precision but estimates published in 2007 in the *International Journal of Environmental Technology and Management* (vol. 7, p. 99) suggest that up to the year 2000 we had burned 110 gigatonnes (Gt) of oil, 60 gigatonnes of oil equivalent (Gtoe) for natural gas and 150 Gtoe for coal. 'Oil equivalent' refers to the amount of oil that contains the same primary energy as a given amount of natural gas or coal, based on standard conversion factors, although these factors can vary according to the source of the fuel, particularly for coal.

By converting the oil equivalent tonnages to actual quantities of the individual fuels we arrive at 110 Gt of oil, 47 Gt of natural gas and 250 Gt of coal. Assuming all the hydrogen in each of the fuels is oxidised to water, one can estimate that oil will generate 140 Gt of water, natural gas 105 Gt, and coal 90 Gt.

Together that's 335 Gt of water, which has a volume of

335 cubic kilometres. The surface area of all Earth's lakes and oceans adds up to about 360 million square kilometres. Spreading the water of combustion evenly over this area would result in rises of about 0.95 millimetres.

The journal also estimates that cumulative oil consumption would increase to 370 Gt for oil, 370 Gtoe for natural gas and 490 Gtoe for coal by the end of this century. This would lead to a total water level rise of about 4 millimetres. With current and projected levels of consumption, this could rise even further.

David Williams
Watson, ACT, Australia

Where there's muck

On the London Underground, I was struck by the black dust that cakes the tunnels. Much of it must be human skin. What is the annual mass of skin cells shed into the underground system? People make a billion journeys on the tube each year. And what other material is the dust composed of?

Richard Fisher
London, UK

Two-thirds of airborne dust in the London Underground is iron oxide, mainly produced by abrasion between train wheels and the track. Most of the rest is volatile matter, while quartz accounts for 1 or 2 per cent. Quartz is a component of brake dust, so we can assume that the proportion of quartz would have been higher in the days before electric motors were introduced to decelerate trains. There are also traces of metals such as chromium, manganese and copper.

The amount of airborne dust increases from dawn,

reaching a peak at around midday. Then it remains fairly constant, with fine dust carried into the system as it is sucked in by air currents from the surface. It then settles overnight.

More details can be found in a 2004 paper by Anthony Seaton and colleagues in *Occupational and Environmental Medicine* (DOI: 10.1136/oem.2004.014332). This research was commissioned by London Underground in response to health-scare stories in the press.

It found that underground dust is coarser than that found above ground. The lower proportion of tinier particles is good news for passengers because bigger particles tend not to penetrate as deeply into the lungs and are cleared out more efficiently.

Skin contributes very little to dust in the London Underground. We each shed about 1.5 grams of skin per day. Journeys by 2.93 million people per day take an average of 44 minutes. If passenger density had been at present levels since 1863, when the network opened, there would be about 7500 tonnes of shed skin. However, spread across the 402-kilometre network, this would equate to a depth of only about 5 millimetres, assuming that tunnels have an average width of 5 metres and that dead skin has the same density as water. Crucially, this also assumes there are no microbes to feed on the detritus.

Mike Follows
Willenhall, West Midlands, UK

To clarify: 44 minutes is about 0.75 of an hour. 0.75 divided by 24 (hours in the day) and multiplied by 1.5 (grams) equals 0.0469 grams in 44 minutes. Therefore the amount of skin left by 2.93 million people is 137,417 grams. Over 149 years (the length of time the network has been open) this amounts to about 7478 tonnes – Ed.

Great circles

What is the longest possible long-haul flight you could take? Or, in other words, which international airports are furthest apart?

Graham Lawton
London, UK

Ignoring the fact that Earth is very slightly oblate and taking its circumference to be 40,000 kilometres, the longest possible great semicircle – that is, a circle whose plane passes through Earth's centre – is 20,000 kilometres. The two major international airports I can find that come closest to this separation are Bogota, Colombia, and Jakarta, Indonesia, which are 19,829 kilometres apart via a great circle.

There is a website for checking this sort of thing. The Great Circle Mapper (gc.kls2.com) requires the International Air Transport Association codes for two or more airports and will plot the great circle routes and distances between them.

Andrew Bristow
Chorley, Lancashire, UK

Given that Wellington, New Zealand, is almost precisely the antipode of Madrid, Spain, and both cities have international airports, this must be a contender. The question then arises, given the slight asphericity of Earth, of which great circle route would be the longest between the two.

Robin Bell
Canberra, ACT, Australia

Presumably your correspondent wishes to know the longest scheduled flight. By private charter one can book planes to their maximum range, and with military in-flight refuelling this can be increased to 40,000 kilometres – say, between the

South African cities of Johannesburg and Pretoria via the poles. No civil aircraft has this capability, though.

The Boeing 777–200LR has a range of 17,445 kilometres but with passengers and international regulations it is less. However, in a record-breaking publicity stunt in 2005, a specially prepared 777–200LR flight with passengers flew eastwards from Hong Kong to London – a distance of 21,601 kilometres.

The longest scheduled flight is by Singapore airlines using an Airbus 340–500 from New York's Newark airport to Singapore. This covers 15,345 kilometres in about 19 hours.

Brian King
Barton on Sea, Hampshire, UK

Both Bermuda and Perth, Australia, have international airports and are nearly precisely antipodes, although I doubt any airline offers a non-stop service between them.

Joe Dellinger
Houston, Texas, US

Sucked under

During coverage in January 2012 of the Costa Concordia *cruise liner disaster off the coast of Italy, I heard some survivors voice concerns about being 'sucked under' if the boat sank. In what conditions would this be likely? How long does the downward force last, and would wearing a life jacket help?*

Bruce Stevens
Hickory, North Carolina, US

There has been much amateurish debunking and misunderstanding of this phenomenon. First-hand evidence from people who have been sucked down is hard to come by, because few survive, but any survivors' accounts make sense in light of the discussion below.

To understand the process, put small, slightly buoyant objects on large weights, let them sink through fluids and observe their behaviour. Start with a pillow or a slab of wood held in the air (air is, of course, a fluid). Scatter slips of paper on top. Most swirl away as the pillow or wood falls, but one or two in the middle will fall with the weighty object.

You can see similar effects with a brick covered in twigs as it sinks in clear water, or try it in slow motion by dropping large ball bearings in a jar of clear detergent containing a scattering of small bubbles. Objects slightly out of line simply swirl, but those caught directly in the wake follow the falling weights like a cyclist slipstreaming a truck.

As a ship goes down, the passengers most at risk are those on the top. Water in a hurry does not query the size of your life jacket; it grips you and down you go. Your best bet, whether in suction or in a rip current, is to swim to one side. You won't have far to go, and then your life jacket can get to work.

Jon Richfield
Somerset West, South Africa

It is quite possible for passengers of a foundering ship to experience the sensation of being sucked under. However, unless the ship is big and sinking quickly – creating a lot of turbulence and releasing a lot of trapped air on its descent – the forces involved are most likely to be small and transient, allowing passengers to swim to safety.

Air escaping from submerged compartments could bubble up through the column of water above the sinking ship. Aeration of water decreases its density and, according to Archimedes' principle, passengers would sink if their weight exceeded the reduced weight of water they displaced. This is why swimmers are less buoyant in the 'white water' of the surf than in the 'blue water' outside the breakers, and why small boats should avoid passing through the white water wake of big ships.

It is thought that bubbles caused by the release of methane gas from methane hydrate deposits beneath the sea floor can sink ships. In 2003, Joseph Monaghan of Monash University in Australia argued that a trawler discovered in a large methane pockmark known as Witch's Hole, about 150 kilometres off the east coast of Scotland, was sunk by a bubble at least as big as the vessel. Bruce Denardo of the Naval Postgraduate School in Monterey, California, tried to disprove the theory by floating a set of small spheres on the surface of a tank of water while feeding bubbling air into the bottom of the tank. The spheres sank. If bubbles can sink ships, the same could happen to people, although people might be able to swim clear of trouble.

While a sinking ship is still just below the surface, passengers could be dragged along in water currents flooding in to displace the escaping air and this helps explain why passengers on different parts of the same sinking ship can have very different experiences.

Charles Lightoller, second officer of the *Titanic*, was twice 'sucked under', carried by water flooding down through ventilators and air shafts. In contrast, chief baker Charles Joughin claimed that he did not even get his hair wet as he stepped off the stern of the *Titanic* while it sank beneath him.

There are other good reasons to stay clear of a sinking ship. For example, when the hospital ship HMHS *Britannic* sank off the coast of Greece in the First World War, a lifeboat full of passengers was caught in the turning propeller as it rose out of the water.

And for passengers left in the water, there is the danger of being struck from below by buoyant objects that break loose from the submerged ship.

Mike Follows
Willenhall, West Midlands, UK

Concertina cars

I was driving down the motorway the other day when the traffic suddenly came to a standstill. After 10 slow minutes of being mostly stationary, the traffic started moving again and was back up to full speed almost immediately. There was no visible reason for the standstill, no accident or junction in sight, and it was slightly later than morning rush hour. Can anyone tell me why traffic bunches up like this for no apparent reason?

Eleanor Harris
London, UK

In open, free-moving traffic, each car is basically autonomous and can travel as fast as its driver wants to go. In denser traffic, there is interaction between vehicles, and if one car slows down then the one behind must also slow down.

When the traffic reaches a certain critical density, a 'shock wave' can spontaneously travel back through it. This is because when one driver brakes gently, the driver behind will choose to slow down more markedly. The effect becomes more pronounced as it works its way backwards. Meanwhile new traffic keeps on arriving at the same rate. With nowhere for it to go, it comes to a grinding halt. The front end of the jam gradually clears, and when cars at the back finally get moving again, the road in front of them is virtually empty, and drivers wonder what the problem was.

Traffic congestion is hard to model because it is very

non-linear and dependent on human reaction times. If drivers had a reaction time of zero and could respond instantly to changes in the flow, an entire highway could start and stop together, like soldiers on parade. But we know from watching cars take off at traffic lights that there is roughly a 1-second delay between each car starting to move. It is probably no coincidence that the critical traffic flow that results in a complete standstill is about one vehicle per second.

Hugh Hunt
Trinity College, University of Cambridge, UK

While travelling towards southern England on the M5 motorway, I saw how a bunch-up in the traffic evolved, from start to finish. This was fascinating to me as a scientist.

Where it occurred, the motorway sloped down for about a mile, then levelled off. Just as I started down the slope, a lorry about a mile ahead indicated it wanted to move from the left-hand lane into the centre lane. This caused traffic behind to start bunching, presumably because drivers eased off the accelerator (I saw no brake lights). This bunching travelled backwards up the slope, like a wave.

When the lorry changed lanes, a second wave started moving up the slope, presumably when drivers braked (their brake lights came on). This second wave travelled faster than the first, merging with it eventually. Where the waves merged, traffic halted. The merged wave brought more traffic to a halt as it moved up the slope past me and disappeared over the crest of the hill.

Once traffic had cleared the point where the lorry switched lanes, it began to speed up and unbunched, creating a third 'acceleration' wave. This, too, travelled up the hill, enabling stopped cars to move and quickly reach the speed they were moving at before the bunching occurred. This third wave also disappeared over the crest behind me.

For anybody who had not seen the lorry's manoeuvre, or for anybody over the crest behind me, there would have been no apparent reason for the stoppage.

Richard Nolan
Cratloe, Clare, Ireland

When a smoothly moving motorway is nearly at full capacity, glitches resulting from a driver changing lanes or braking will cause cars behind to brake, creating a mass of slow-moving or stopped cars. If the number of cars that joins the back of this mass matches or exceeds the number of cars leaving over the same period, the mass will persist and even grow. It may do so for extended periods, dissipating only when traffic flow decreases for long enough that more cars leave than join, slowly dissipating the mass.

A similar phenomenon occurs in data networks such as the internet: during periods of congestion, queues of data packets build up quickly, persist, and are slow to dissipate. I described the precise mechanism for this phenomenon in my PhD dissertation.

Srinivasan Keshav
School of Computer Science
University of Waterloo
Ontario, Canada

To kill these waves in dense highway traffic, leave six to eight car lengths to the next vehicle, and as soon as you can see the brake lights of cars further along go on, take your foot off the accelerator and coast. If the car ahead of you regains speed, you can easily catch up, so avoiding a general slowdown. The cars behind will have slowed down only slightly while you coasted, so they can also speed up when you do.

The idea sounds good on paper, but unfortunately other

drivers may not be able to comply as the usual habit is to keep too close to the car ahead.

Don L. Jewett
University of California, San Francisco, US

Lane changer

In multi-lane traffic jams on the motorway, I often seem to move into the faster-moving lane just as it becomes stationary and the lane of traffic I just left starts moving. Assuming that other drivers experience the same, or are not deliberately trying to stymie me, what is the best strategy for getting through multi-lane traffic jams as quickly as possible?

Peter Slessenger
Reading, Berkshire, UK

Ten years ago, I had a daily commute along an 85-kilometre stretch of London's M25 motorway. After a few weeks, I began to notice that in congestion I'd overtake the same lorry again and again. On that basis, I decided that as soon as traffic started to get stuck, I'd go into the 'slower' (non-overtaking) lane. I soon discovered that I'd normally get through the congestion before the cars that either stayed in the overtaking lanes or attempted multiple lane-shifting.

More recently, I commuted several times a week on the autobahn between Basel and Vevey in Switzerland. I observed the same phenomenon. Presumably the perception that other lanes are moving more quickly is partly psychological – the nearside lane is full of lorries going slowly, so it must seem slower when you are in it. Of course, there can be local variations according to conditions and the cause of the hold-up.

My advice is to get into the nearside lane as soon as

congestion appears, and stay in it until the traffic flows again. Be aware, however, that overtaking on the nearside, except in slow-moving traffic, is against the law in many countries.

As a rule, I found that for a journey of about 160 kilometres, driving in this way (and, in general, driving defensively) made little or no difference to the total journey time compared with driving aggressively. However, driving defensively, I'd arrive considerably less stressed.

Matt Billingham
Binningen, Switzerland

The BBC ran an experiment on just this question many years ago. Two cars set off on identical long journeys on congested motorways on the first day of the summer holiday getaway. One car stayed in the nearside lane, the other tried swapping lanes as described in the question. It made no difference to the overall journey time, so it seems you might as well stay in the nearside lane.

Pam Lunn
Kenilworth, Warwickshire, UK

Australian television came up with a slightly different result – Ed.

A few years ago, an Australian science television programme did an experiment during peak hour to find the quickest way to drive through multi-lane traffic jams. Two cars started at the same place and time, and went to the same destination following the same route. One driver changed lanes whenever he thought another lane was moving faster, while the other driver stayed in one lane and didn't worry about how long it took to reach the destination.

The result was that the lane changer got to the destination more quickly – but only just (something like 3 minutes earlier over a 30-minute trip). However, in doing so, the lane changer used significantly more fuel (about 25 per cent more), and reported feeling much more stressed and aggressive at the end of the drive.

The conclusion was that it was better for the environment and for your health and stress levels to chill out and stay in one lane – and it didn't make much difference to your overall travel time.

Adam Friederich
O'Connor, ACT, Australia

I have a 160-kilometre commute to work and have experimented with various options in an attempt to beat the traffic. The best one? Get a motorcycle.

Daz Loczy
Nottingham, UK

Mountain climbing

I've been watching the Tour de France and the stages in the mountains are fascinating. It helps the best riders if they have teammates riding up the steep slopes in front of them, but surely they are travelling too slowly for this help to come in the form of a slipstream, like on the flat, fast stages. So why is it easier to follow another rider up a mountain slope? Is the effect purely psychological?

Nicola Dorset
King's Lynn, Norfolk, UK

The effect of slipstreaming at the low speeds involved when mountain climbing is small, at best, and perhaps not significant. However, the psychological effect of having a teammate's support is hard to quantify, but very real.

This goes hand in hand with the physiological benefits of team support. A cyclist like former Tour de France champion Bradley Wiggins can climb mountains very fast at a steady pace. He is not, however, a true climber like the late Italian cyclist Marco Pantani, who had the ability to accelerate rapidly while climbing. For Wiggins, having his teammates set a fast pace means that he can repel the suppressed attacks from the Pantani-like climbers, who can struggle to match the constant high speed. At the same time, this tires them and reduces their ability to attack with rapid acceleration later on. Near the end of a stage Wiggins is able to increase the effort (as is his successor as Tour champion Chris Froome) to a speed that most other riders are unable to match.

Wiggins is an outstanding time-trialist, able to maintain a steady but high level of speed over long distances, and the tactics he used in the Tour de France effectively turned the mountain stages into uphill time trials, which played to his strengths.

Michael Newman
Durrington, Wiltshire, UK

As long as there is air movement, caused either by local winds or a rider's speed, then there is at least some benefit in having the slipstream shelter of a teammate.

Additionally, the teammate climbing the mountain with his team leader will carry food and drinks, lightening the load that the team leader has to carry himself. This helper can be replaced and rested the next day, but the leader cannot.

The helper will also give up his wheels or even his bike should the leader encounter a problem, and this prevents the leader losing time on his rivals in these circumstances. There is, therefore, a degree of moral and mechanical support in having a teammate ride with you.

Finally, speaking from my own long personal experience,

there is considerable psychological advantage in following a wheel. This is especially noticeable when you are hemmed in by other riders behind and to the side of you, which gives you little choice. Remember that the leader is the one setting the speed at which his teammate rides, by shouting out a certain tempo or power output. The pace is therefore one he knows he can sustain. All he has to do is follow the wheel in front.

Martyn Ellis
Bicester, Oxfordshire, UK

Train brain drain

If I don't know the time of the next half-hourly train to depart from my local station, is there any point in running to the station? My wife tells me it won't make any difference to whether I catch the next train or not, but I insist it will. I must admit, my journey time door-to-door does not seem to be affected by the speed of my approach to the station.

Jeremy Browne
Elm Park, Essex, UK

If you arrive at the railway station at some random time, regardless of how fast you travelled to the station, you will have to wait an average of 15 minutes and a maximum of 30 minutes for the next train. So running will not make any difference to your average waiting time at the station.

However, if you arrive 1 minute earlier by running than by walking, then on one trip in 30 you will catch a train that you would have missed if you had walked, therefore arriving at your destination 30 minutes earlier.

On the other 29 out of 30 times that you don't catch an earlier train, you will have to wait 1 minute longer on the platform than if you had just walked there. On these occasions the trip will probably seem longer, although the actual door-to-door time remains unchanged.

You save 30 minutes once in 30 trips, so on average the 1 minute saved by running saves 1 minute door-to-door. If you save more than 1 minute by running, you've got a better chance of catching the train and saving 30 minutes more often. The average time saved always matches the time saved by running.

Brian Horton
West Launceston, Tasmania, Australia

If your run gets you to the station 3 minutes earlier, then one day in 10 you will get to work half an hour earlier. So your wife is wrong. But perhaps she meant that the average waiting time at the station will be unaffected. This is true: nine days out of 10 you will wait 3 minutes longer, exactly balancing the one day in 10 that you save 27 minutes.

None of this explains the fundamental question underlying this paradox: why not look at the train timetable? If you are uncomfortable with smartphone apps, timetables still exist in old-fashioned paper form.

Ian Gent and Judith Underwood
Cupar, Fife, UK

Faster than the wind

In the 2013 America's Cup races, the boats were frequently sailing at more than twice the wind speed. For instance, in the final race the wind speed was given as about 20 knots, while the boats were travelling at well above 40 knots. How is this possible?

Chris Evans
Earby, Lancashire, UK

A sailing boat is not a mere wind-blown object, like a leaf or hot-air balloon. The sail is cut so that it assumes a shape similar to an aircraft's wing. The wind travels faster along the convex side, creating lift. It is fastest with the wind blowing across the boat's side. But this force also generates an opposite reaction, which acts as drag on the boat's keel – the fin-shaped part extending down into the water from the hull that provides stability.

In addition, as a boat sails, its own speed adds to the wind's speed. This is similar to running in still air when you can feel the 'wind' on your face.

This 'apparent wind' creates further lift from the sail and the speed of the boat increases. Boats with keels have a limited speed due to the drag this feature creates. Ice yachts don't have this constraint and can sail much faster, until aerodynamic drag limits their speed.

The disputed world speed record for an ice boat is about 230 kilometres an hour, but over 88 kilometres an hour is commonplace.

John Davies
Lancaster, UK

The sails of a boat are curved in a similar way to an aircraft wing. They are adjusted by the crew to ensure that the air flows evenly over them. This maximises the extraction

of energy from the wind, which is used to drive the boat forwards.

As well as the true wind, a boat experiences an 'apparent wind' caused by its own motion, and it is this effect that enables boats to sail faster than the true wind. In the America's Cup example, simple vector analysis shows that a boat travelling at 74 kilometres an hour, broadside on to a 37 km/h wind, will actually be sailing in an 83 km/h apparent wind.

The bigger the sail area, the more energy can be captured, but the sails have to be kept as vertical as possible. Large sails create a huge turning momentum that can rapidly capsize the boat. To counter this tendency, some boats use a large weighted keel; dinghies and catamarans rely on the weight of the crew. But increasing weight tends to lower the boat in the water, and this increases drag.

Most boats have a limitation on their speed, known as the hull speed, which is roughly 1.34 times the square root of their length in feet. Increasing the power in such cases, through engines or sail area, only results in a bigger bow wave and no increase in speed.

However, clever design coupled with sufficient power can induce a phenomenon known as planing. In this case the hull can sit higher in the water and even sometimes briefly come clear of the surface as the boat skims along, much faster than its hull speed. Racing power boats often exhibit this phenomenon.

A further development of this is the hydrofoil, where the whole hull lifts out of the water with even greater drag reduction. Long, thin, wave-piercing hulls also reduce drag but are not very stable. If you place two of these in parallel and separate them, with the mast and sails in between, you have created a stable and fast catamaran. Now just add the biggest sails you dare, foils that lift the hull out of the water, plus Ben Ainslie on the helm and you have designed a winning entry for the next America's Cup.

James Stafford
London, UK

Silent fleet

One argument for continuing the UK's nuclear submarine fleet is that they are undetectable. Is that still true? And is it likely to be correct when replacement vessels are ready?

Barry Cash
Bristol, UK

This is a controversial and often misunderstood issue. While nuclear-powered submarines have advantages in speed, range and submersion time, they are far from undetectable. Surprisingly, modern diesel submarines are often far quieter than their nuclear-powered counterparts.

The problem lies in the power plant itself. A nuclear reactor cannot be turned off and on like a diesel engine can. Although a nuclear submarine can rapidly stop its propeller, it wouldn't shut down the reactor because of its long start-up and shut-down times.

When it is running, the reactor emits an acoustic and thermal signature which can be picked up by modern detection equipment. In contrast, diesel subs can shut down their entire propulsion system, lying almost completely silent on the bottom of the ocean.

Oxygen-free propulsion and battery technology in diesel submarines have improved vastly since the inception of the nuclear submarine, increasingly closing the gap between nuclear boats and their cheaper conventional counterparts.

Next-generation submarines promise to be even quieter, using hydrogen-cell propulsion and nanotechnological acoustic barriers. These vessels would be virtually undetectable using current anti-submarine military technology.

However, no vessel can ever be truly 'undetectable'. As submarine technology advances, so too will the techniques and technology used to find them.

Benjamin English
First lieutenant, US Marine Corps
Pensacola, Florida, US

Germany's U-480 submarine used in the Second World War was arguably the first stealth submarine. It was difficult to detect using sonar because it was coated in rubber that contained air pockets. The cat-and-mouse game has continued unabated ever since.

Submarines usually lurk on continental shelves, where they are difficult to detect against the clutter of sonar echoes off the surrounding underwater topography. Skilled submariners also position their boats amid water layers called thermoclines, where temperature changes abruptly, refracting sonar away.

Propellers are deliberately not spun so quickly that cavitation bubbles are created because when these bubbles collapse they create a tell-tale acoustic signature. And, during 'silent running', the crew will make minimal noise.

Meanwhile, modern techniques reduce acoustic signatures. Apart from quiet propulsion systems and running on electricity by direct current, deck structures are isolated from hulls with vibration-absorbing mountings and hulls have coatings that eliminate echo.

The joint Italian and German Type 212A class submarines promise to be quieter still. Powered by hydrogen fuel cells, they will radiate virtually no heat. They also use a non-magnetic construction, and it is claimed they will be nearly impossible to detect. Nano-coatings on the hull, designed to reduce fouling caused by organisms – and the associated turbulence that comes with it – might also guide sonar smoothly around the outside of the hull, making science-fiction's cloaking device a possibility.

Mike Follows
Sutton Coldfield, West Midlands, UK

13 The Rest

Stop talking

What parameters, if any, limit the number of different words available to us in English (or any other language)? Are we near to running out of words?

Jonathan Cope
London, UK

The existence of polyglots suggests that the average person is far from running out of storage space in the brain. And while there are major structural constraints on word numbers, they too leave lots of room for expansion.

First, there is some effect from the number of distinct sounds (phonemes) in a language. Languages with fewer phonemes have some tendency to have longer words.

English has about 40 phonemes and many short words, while Hawaiian, with only 13 phonemes, has many words of three or four syllables. Any language can increase the potential size of its vocabulary by using longer words.

Much more limiting than the number of phonemes are a language's phonotactic patterns – the constraints on possible sequences of phonemes. In English, words can begin with sequences like 'sp' or 'st', but in Spanish they cannot hence the vowel at the beginning of Español. In Greek, words can begin with sequences like 'pn' or 'ps', but in English this is not permissible, so we pronounce Greek loan words like pneumatic and psychology without the 'p'.

Even so, we are a long way from using up all the permitted shapes: AIDS, fax and ROM are all recent additions, but 'snizz', 'whask' and literally thousands of other possible English sequences are all unused.

D. Ladd
Department of Linguistics
Edinburgh University, UK

Assuming that we strictly limit ourselves to only consonant–vowel–consonant forms and do not include such extras as tones and stress, the following provides a generous lower boundary to the number of words available in English.

There are more than 50 possible initial consonants (including combinations such as 'tr' and 'sk'). There are more than 10 distinct vowel sounds. Consider: rad, raid, red, rid, ride, rude, rod, reed, road, and add in non-words such as roid (void) and rould (should).

There are more than 40 terminal consonants (including 'rt' and 'lk'). This means that there are in excess of 20,000 (50 x 10 x 40) single syllables. If we limit ourselves to using only two syllables per word we would still have more than 400 million words to play with.

Francis Glassborow
Oxford, UK

Numerical differences

On a trip to Saudi Arabia I expected to find 'Arabic' numerals in use (as we call them in the western hemisphere). Arabian numerals are entirely different, so how did ours become known as Arabic?

Jerry Kluza
Brookfield, Wisconsin, US

Both sets of numerals originally come from India. Although the numerals used by Europeans, Americans, and much of the international community are referred to as Arabic numerals they were not originally created by the Arabs. This misnomer originated in the 9th century when a manuscript on arithmetic which had been written in India was translated soon after into Arabic. Merchants then carried this book to Europe where it was subsequently translated into Latin.

Because the source for the Latin translation was an Arabic text, the numerals were falsely ascribed to the Arabs. This is where the confusion arose: the numerals came from India and are not Arabic at all.

The present set of Arabic numerals evolved over the early centuries but has changed little since the advent of printed books in around 1445. Curiously the numerals 4, 5, 6 and 7 underwent the most changes from the original script. The numerals used in Arabic countries (and Iran) have undergone minor changes from the original Hindu manuscript.

Paul Marselian
San Diego, California, US

Indians made two crucial advances in number systems. They introduced place values and a special symbol for zero.

The value of these innovations can readily be experienced by anyone trying to do simple arithmetic with non-place-value systems, like Roman numerals. Place-value numeration allows the mathematician to arrange addition in columns and to carry over numbers from the ones to the tens, or from the tens to the hundreds columns. This is very difficult to do with Roman numerals.

It is interesting that all Indic scripts are written from left to right (like English), and so are the numbers. Arabic writes words from right to left. Yet its numbers, being borrowed from those used in India, are written from left to right. Many speakers of Arabic are unaware of the reason why their numbers are written in the opposite way to their words, even though they may use them every day.

William Wolf
Chicago, Illinois, US

We did indeed get our numerals from the Arabs, who had adopted them originally from India. In Arabic they are therefore called *al-arqám al-hindíya,* which translates as 'the Indian numerals'.

Philip Stewart
Oxford, UK

The numerals that we use today are called Arabic because Europeans adopted them from the Arabs. However, they are Indian in origin.

According to popular tradition the Arabic numerals were invented soon after the start of the 6th century by the Indian astronomer Aryabhata. They made up the first fully positional numeral system, in which the numeral 1 could stand for 'one', 'a hundred' or 'a thousand', depending on its position in the number.

It was not the first use of a symbol for zero, however. This occurs in late Babylonian numerals. Aryabhata is said to have been inspired by the sand abacus, a device for multiplying numbers by making marks in a grid drawn on sand, in which the position of the marks is significant.

The Arabs brought Indian scholars to Baghdad in 771 to teach them how to use the new numerals, which they originally called *huruf al-ghubar*, or sand letters, after the sand grid mentioned above.

Europeans started to use them widely after 1202, when

Leonardo of Pisa explained and detailed their use in his *Book of the Abacus*.

Ralph Hancock
London, UK

In 1299, the city of Florence in Italy issued a decree prohibiting the commercial use of Arabic numerals because they were easily falsified. For example, a '0' could easily be changed to a '6'.

Ron Webb,
Macclesfield, Cheshire, UK

The precise glyphs that are used for the digits have varied considerably over the centuries. Those used in Europe were stabilised by the selection William Caxton used for his printing press.

Prior to that there was considerable variation, particularly for the glyphs that were used for 4 and 5. For a long time the glyph we now use for four would have been considered to represent five.

It is not surprising that the actual glyphs used in Arabic-speaking countries are very different from those that we use now, and these differences can be very confusing, especially to the unwary.

The traffic speed limit on the bridge running between Khartoum and Omdurman in Sudan used to be posted in both English and Arabic. The limit was 10 miles per hour, but apprehended drivers frequently defended themselves on the grounds that they had never exceeded the 15 miles per hour that they had seen on the signpost. This is because the glyph we use for zero is very similar to that used in Arabic for five.

To add to the confusion, the Arabic glyph for zero is also very close to the raised point that we traditionally use as a decimal point.

Francis Glassborow
Oxford, UK

Roman division

How did the Romans express fractions?

John Chappell
Grassington, North Yorkshire, UK

In scientific work, which was always written in Greek, the Romans used sexagesimal fractions, like those used for angles and time, expressed by Greek literal numbers and positional notation.

For everyday uses, common fractions were always used, often spelled out, such as *tribus duas partes* ('with thirds two parts') for $\frac{2}{3}$. The solidus (the line in a fraction between the two numerals), or any implied division, was never used. However, abbreviations were often used if necessary. The most common were S or SK (*semisque*, or 'and a half') for $\frac{1}{2}$, and T or TK for $\frac{1}{3}$; $1\frac{1}{2}$ could be written ISK. A symbol for the sestertius (a unit of money worth $2\frac{1}{2}$ bronze asses), was IIS, later written HS. F, Z or FZ was used for $\frac{2}{3}$ while $\frac{1}{4}$ was represented by the usual modern division symbol ÷, or :- -, or G, or a backward C. An overhead horizontal bar meant either $\frac{1}{12}$ or $\frac{1}{16}$, and could be combined with dots and other symbols. Common fractions were often expressed as sums of simpler fractions, for example $\frac{9}{16} = \frac{1}{2} + \frac{1}{16}$, written S- -.

A good place to see such fractions is *Book X* of Vitruvius's *De Architectura*, where they're used in connection with the construction of military machines. However, ignorant medieval (and modern) copyists have corrupted many fractions in existing texts because they did not understand the unfamiliar symbols. There are also a few examples of fractions in inscriptions but evidence is scanty and, as classical scholars usually pay little attention to numbers, science or mathematics, the usual reference works provide no enlightenment.

Roman practical calculations were done on the abacus,

which used a decimal notation for whole numbers, but not for fractions (which were usually monetary fractions), and these had special columns of beads.

Incidentally, the notation for large numbers was different from that used in today's Roman numerals, and more convenient. Except for I, V and X, Roman numerals were not the usual alphabetical letters we see today, but symbols derived from Etruscan and other sources. L, C, D and M evolved from these, but many of the more unusual symbols remained for fractions. Even Greek letters are sometimes found. X was also used for $\frac{1}{16}$.

So, except for the simpler fractions, there was little standardisation. Roman engineers relied more on analogue and graphical methods of calculation than we do today, and numerical calculations were avoided as far as possible.

Jim Calvert
Exeter, Devon, UK

The Egyptians used fractions and evidence survives in a few papyri, notably the Rhind papyrus, bought in Luxor in 1858 by Henry Rhind – now in the British Museum. It dates from 1650 BC and was probably copied from a text written two centuries earlier. It shows methods of calculation, problems, exercises and puzzles involving fractions. There was an Egyptian symbol denoting fractions and the numerator was always 1, except for $\frac{2}{3}$ and $\frac{3}{4}$. Even these could be derived by subtracting $\frac{1}{3}$ and $\frac{1}{4}$ from the whole. Intermediate values could be expressed as the sum of unit fractions, for example $\frac{3}{5} = \frac{1}{3} + \frac{1}{5} + \frac{1}{15}$.

Abstract fractions, as we know them today, were probably invented by Hindu mathematicians in about AD 500, and the horizontal bar (the solidus) was introduced by the Arabs around 1200.

John Richmond
Worcester Park, Surrey, UK

Candle in the wind

My birthday cake had candles on it that I couldn't blow out. How do these work?

Luke Mannion
Brisbane, Queensland, Australia

The short answer is that they have magnesium incorporated into the wick.

For a flame to exist, it needs oxygen, fuel and enough heat to keep combustion going once it starts. These requirements are sometimes represented graphically as the three sides of a 'fire triangle'.

When you blow out an ordinary candle, you extinguish the flame by removing heat. You can still see the fuel – the wax smoke, often paraffin vapour – coming from the wick, but the ember in the wick does not supply enough heat to reignite it, and so it will eventually go out. But when the wick has magnesium powder in it, as was the case with the candles on your correspondent's birthday cake, the ember is able to ignite the magnesium. This element catches fire at relatively low temperatures, below 500 °C, and this is enough to reignite the smoky fuel. The flame itself then burns at around 3000 °C.

If you look closely at one of these candles after you have apparently blown it out, you can see the smouldering wick emitting little sparks of burning magnesium before the candle relights.

Oh, and a belated happy birthday to your correspondent.

David Muir
Science Department
Portobello High School
Edinburgh, UK

Spin doctors

I have been watching TV coverage of international cricket matches this summer. The channel uses a new device that measures how quickly bowlers spin the ball. The results seem amazing. Even slow spinners manage more than 1000 revolutions per minute, and some spin the ball faster than 3000 rpm, which equates to more than 50 revolutions per second. Is this correct? And if it is, how do they spin it so quickly?

David Mushens
London, UK

Doppler radar techniques have been measuring the rotation of balls in sport for years. In cricket, bowlers are classified as either finger-spinners or wrist-spinners, so named to reflect how the ball is released from the hand.

Finger-spinners keep the ball in contact with their forefinger for a short time after release, whereas wrist-spinners use their little finger. This 'wrapping' of the ball around the finger gives it spin. Many different spin speeds and trajectories can be achieved when you add the effects of left or right-handed bowling, differences in hand rotation and the position of the ball's seam in relation to the hand and the spin axis.

John Davies
Lancaster, UK

If bowlers imparted spin merely by simply rotating the ball as though they were turning a dial, such rotational speeds would be impossible. But that is not how it is done.

Just as the rotation rate of a ball determines the speed at which it travels along a surface, the speed at which a surface rides over such an object controls its rotation rate.

When bowlers launch the ball, it stays in contact with and rides over the palm side of the hand, which imparts spin via friction. International match cricket balls have a maximum circumference of 229 millimetres.

To rotate at 3000 revolutions per minute, the ball has to traverse the hand at a speed of 11 metres per second. For comparison, a hand's velocity during a karate chop is between 10 and 14 metres per second. The bowler can boost this spin by rotating their forearm about its long axis while throwing.

In truth, the spin speeds seem surprising because the zeros mount up when we convert units – seconds to minutes, for example – and because of our difficulty relating linear to rotational speeds.

Len Winokur
Leeds, UK

Aerial combat

How do the people monitoring television viewing figures know that the numbers they release are accurate? They often total in their millions, but nobody has ever asked me which shows I have watched.

Stephanie Hilden
Portadown, Armagh, UK

My wife and I are members of the large panel of viewers on which UK viewing figures for television are based. The panel comprises members of the public selected to be representative of the population. This reflects the geographical distribution, household size, gender, age and so on in the population as a whole.

All participating households have a box next to each TV which records which station is being watched. We have a special remote control which we use to log in individually

whenever we watch. This remote also has facilities to record guests who may be present together with their approximate age, a 'no viewers' button and an 'on holiday' button.

The log of our data is sent automatically to the monitoring organisation in the early hours of each morning. We notify them of any changes, such as a new or extra television, and periodically we receive a phone call to confirm that all is working correctly. We do not get paid but are offered points that can be exchanged for vouchers for goods. We receive the same points each month – we do not get more for watching more TV.

Martin Trevelyan-Jones
Conwy, Clwyd, UK

Viewing figures are collated using a variety of methods depending on the country; in the UK, the most-quoted figures are from the Broadcasters' Audience Research Board. Generally, they are calculated by surveying a representative sample of households and then extrapolating the results to represent the whole country.

There are other methods used by individual broadcasters. For example, cable and satellite broadcasters can monitor which of their channels are being viewed and create aggregate figures for viewers. Programme makers, advertisers and broadcasters also use methods such as street and online surveys to gain information on viewing habits.

Some researchers are testing new systems using cameras that recognise when individuals are in a room and detect how often they look at the screen. In this way they can assess how much attention is being paid to various programmes or adverts.

John Thompson
London, UK

Index